Brot
für die Welt

WAU ECOLOGY INSTITUTE HANDBOOK No. 10

SUBSISTENCE AGRICULTURE IMPROVEMENT

Manual for the Humid Tropics

Edited by
Friedhelm Goeltenb

ISBN 978-9980-945-76-1

Subsistence Agriculture Improvement: Manual for the Humid Tropics (Wau Ecology Handbook, 10)
Edited by Friedhelm Goeltenboth

Contributing Authors: V. E. Dalpadado, F. Goeltenboth, N. H. S. Howcroft, B. Louman, G. Nalu, F. Samana, B. Siki

Illustration: Sally Shaw

Previously co-published by:
Wau Ecology Institute
P. O. Box 77
Wau
Papua New Guinea

and by:
Margraf Scientific Publishers

Previous publication was supported by Bread for the World, Stuttgart.

Reprinted with permission in 2012 by:
University of Papua New Guinea Press & Bookshop
P. O. Box 413
UNIVERSITY PO
NCD
Papua New Guinea
Tel: 675 + 326 7281 or 326 7675
Email: upngbooks@gmail.com

ACKNOWLEDGEMENTS

The editor and the authors gratefully acknowledge the help received by the WAU ECOLOGY INSTITUTE and its director Mr. Harry Sakulas MSc., the Project Planning Committee chaired by Prof. Dr. E. Balasubramaniam, Miss Sally Shaw (USA) for her excellent drawings illustrating this manual, Mrs. Ellen Woodley (Canada) for her substantial help in correcting the manuscript and Bread for the World, Stuttgart, West Germany for providing financial assistance.

THE WAU ECOLOGY INSTITUTE

P.O. BOX 77

WAU

PAPUA NEW GUINEA

The Wau Ecology Institute is located in the mid-mountain area at Wau (Morobe) in Papua New Guinea.

The Institute was established in 1972 as a non-profit, non-governmental corporation dedicated to the study and education of ecology and conservation, rural development and environment protection in Papua New Guinea.

Located at about 1000 meters altitude on the lower slopes of Mount Kaindi, the Institute has access to a remarkable range of environments at various altitudes.

The summit of Mt. Kaindi (2350 m) is accessible by road, and the extensive pristine forests provide abundant wilderness tracts for survey and study of the biota.

The Institute has facilities for short-term and long-term visitors. Short-term guests may stay in the hostel or in rental bungalows. Long-term researchers and visitors may lease Institute housing and laboratory space. In addition, the Institute supports a regional research library, arboretum, herbarium, small zoo, medicinal plants demonstration garden and preserved collections of vertebrates and invertebrates. Research facilities include dehumified laboratory space, microscopes, preserving equipment and assorted research supplies.

Visitors, researchers, naturalists and bird-watchers are welcome to visit and stay at the Institute. Because of limited space, all are adviced to make bookings in advance, and to make full inquiries about the availability of specific services, supplies or transportation:

The Director
Wau Ecology Institute
P.O. Box 77
Wau, Papua New Guinea
Phone 44-6341

AUTHORS AND THEIR CONTRIBUTIONS

	Present address
Victor E. Dalpadado	Box 831 Lae Papua New Guinea (Application of work study techniques, points about extension methods and nutrition improvement in subsistence farming)
Dr. Friedhelm Goelthenboth	Hauptstr. 9 7401 Nehren West Germany (Introduction Notes on subsistence agriculture and the subsistence agriculture improvement project Background information about the mid-mountain biogeography of Papua New Guinea Appropriate tools for the subsistence garden Subsistence agriculture and aspects of ecology and conservation in Papua New Guinea Aspects of subsistence agriculture and education in Papua New Guinea)
N.H.S. Howcroft	Forest Research Station P.O. Box 134 Bulolo Papua New Guinea (Nitrogen fixing trees)

Bas Louman
Thorbeckelaan 4
5463 BM Veghe
The Netherlands

(Appropriate tools for the subsistence garden)

Fungke Samana
Morobe Women's Association
P.O. Box 1327
Lae
Papua New Guinea

(Women in agriculture)

Beka Siki
Box 697
Lae
Papua New Guinea

(Subsistence agriculture and soil conservation

Agroforestry – a short overview

Simple organic plant protection)

Goaro Nalu
c/o DPJ
Wau
Papua New Guinea

Practical advisor

FOREWORD

Subsistence agriculture maintains 80% of PNG's 3 million people. Subsistence agriculture is the major pre-occupation of the bulk of the rural population. It is their way of life.

Whilst, the Nation grapples with the concerns of the state, namely the running of the bureaucracy, and investments involving high finance and trade in need of cash, the villagers grapple with the basic needs of the family — food and shelter.

Yet, when one, assesses the imports of basic food items, canned or otherwise, one wonders about what our Government's planning is up to, and where our priorities lie.

Subsistence agriculture gets the least attention, or none at all by Government agricultural extension services.

Thousands of kina was spent year in and year out on *didiman* cars, fuels and maintenance costs, not to mention payments for travel allowances, boot allowances, accommodation allowances, etc. etc. on extension programmes that concentrate on cash-crops and cattle that promotes "bisnis" and so-called development despite the escalation of import bills on basic food items.

Subsistence agriculture in the eyes of the officialdom is not an "economy."

With this attitudinal background, any effort in the promotion and improvement of subsistence agriculture by way of improving agricultural techniques and methods, is of great importance to the bulk of the population.

This manual on subsistence agriculture is the end result of consistency and persistence in the development of appropriate and improved techniques in improving subsistence agriculture production — to make justice to the people of the land in the villages.

My sincere gratitude to those hardworking individuals and extensionists formerly attached to the Wau Ecology *SAP* program (Subsistence Agriculture Programme), to the contributors to this manual and the Editor Dr. F. Goeltenboth, for the collation and analysis of experiences gained so far, to help the people of Morobe and PNG to further help themselves.

May I commend this manual to all those who are concerned with the practical approach to improving the Melanesian Way of life through subsistence agriculture.

Utula U. Samana
Premier of Morobe Province
Papua New Guinea

FOREWORD

All over Papua New Guinea forests are moving further away up the sides of mountains. The shifting cultivation needs more and more agricultural lands.

The semi-nomadic subsistence agriculture techniques of "slash and burn" are ill-adapted to meet the needs of Papua New Guinea's rapidly expanding agrarian societies.

The "slash and burn" system destroys the forests and as a result the topsoil washes away and only kunai will grow there.

Therefore the Subsistence Agriculture Project of The Wau Ecology Institute developed appropriate techniques for the training of all peoples concerned in an improved subsistence agriculture program showing how to grow quantitatively and qualitatively better food in a site-stable garden.

All the methods used are put together in this manual hopefully making it a resource for all those concerned with the improvement of the life situation of the subsistence farms.

Professor E. Balasubramaniam
Chairman, Subsistence Agriculture Project.

30th October 1984.

FOREWORD (2nd issue)

The steady increasing acceptance of the knowledge that the subsistence farmers are of fundamental importance for a desirable development in a given environment leads to demand of information, how subsistence agriculture could be improved.

By combining the necessity of conservation of the environment with the need for more and healthier food crops, the unique fauna and flora could be protected as "a heritage, worth more than gold", as the most Honourable Sir Paulias Matane stated.

It is felt, that an increasing number of concerned people, multiplicators and officials have realized the necessity to combine the efforts of villagers, scientists and practicioners to disseminate available information to promote the acceptance of environmentally safe practices of subsistence improvement.

After only four years since the publication of the first issue of this book, this second issue is needed. The improved version of the book should help to disseminate the information, how the situation of the subsistence farmers in a montane tropical area could be promoted. It is also hoped, that it will be an example for others to do similar work for other regions of the tropical zone, an environment already under an increasing stress.

Tübingen, August 1989

Dr. Friedhelm Göltenboth
(Editor)

CONTENTS

Introduction

Papua New Guinea is experiencing a rapid depletion of its forest resources through timbering, agricultural development and traditional agricultural practises (Fig. 1) in a new context of local overpopulation (Fig. 2).

Traditional methods involve the cutting of forests for temporary gardens and repeating this process when the soil becomes depleted (Fig. 3). With increasing population growth this method can result in the elimination of the primary forest and create a depauparate environment, such as a grassland, which has nutrient depleted soils.

On the other hand primary forests may be transformed to monocultures for export commodities without regard to the villager's needs and the native flora and fauna. This is particularly true for areas where big timber projects are set up.

Since most of the people in the country still live in villages, some far away from roads and airstrips, subsistence agriculture is very important to their existence. Steps have to be taken through extension education to modify the practises that result in soil loss and nutrient depletion in subsistence agriculture throughout Papua New Guinea. The traditional custom of swidden or slash-burn agriculture was adequate under the earlier low population situation, but with current rapid population growth it threatens the forests, especially on slopes (Gressitt, 1982). People of the forested and sparsely inhabited areas practise long fallow. In the anthropogenic grasslands the people fallow their gardens for shorter periods and substitute various soil-enriching techniques for soil regeneration.

A great variety of horticultural practises and strains of food plants have accumulated in Papua New Guinea over the centuries (Fig. 4). They are closely adjusted to local farming and environmental conditions, thus adaptive agricultural structures are all too easily destroyed by population pressure or cash-cropping (Cox, 1979).

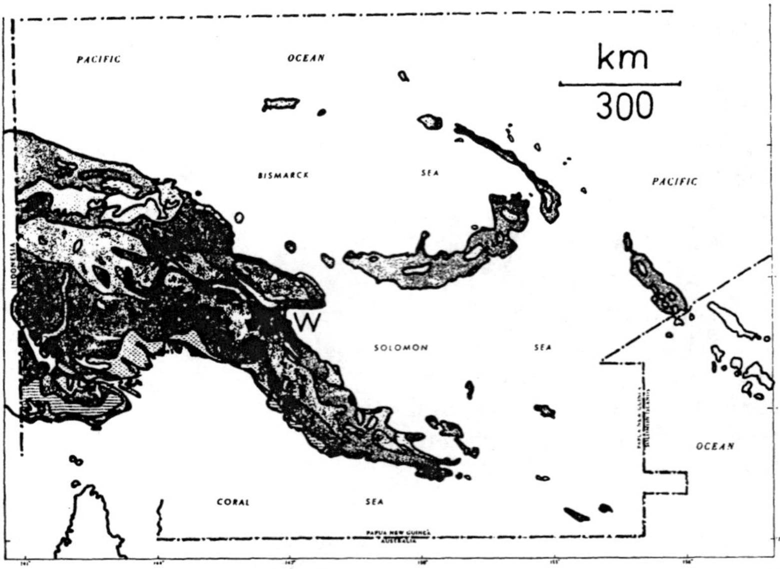

Fig. 1 Map of the vegetation distribution in Papua New Guinea (PNG Atlas, 1978, changed) W Wau

- Lowland Forest
- Mountain Forest
- Grassland
- Savanna
- Wooded fresh water swamp
- Mangroves
- Gardens and grassland
- Herbaceous freshwater swamp

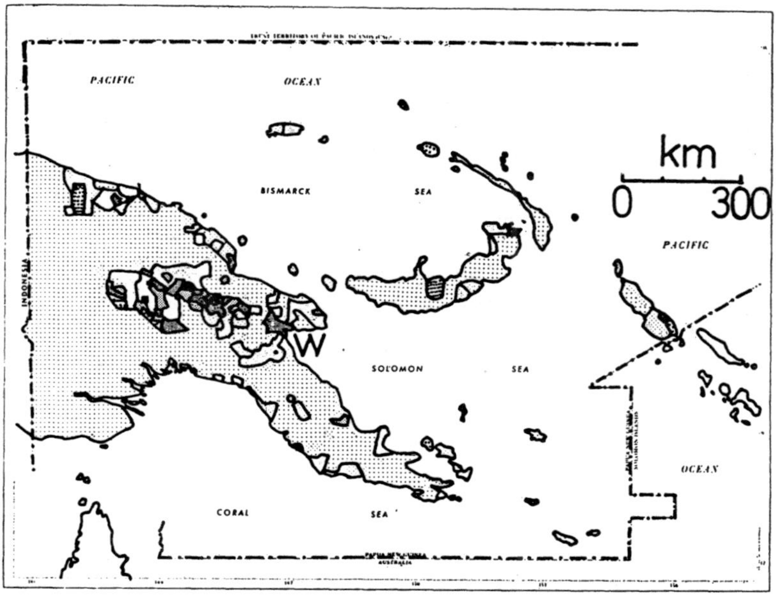

Fig. 2 Population distribution in Papua New Guinea. Stand 1978. (PNG Altas, 1978, changed) W Wau

Persons per km²

- 0–4.99
- 5–19.99
- 20–29.89
- 30–49.59
- 50–69.99
- 70–89.99
- 90+

Recognizing the impact of extensive farming systems on tropical forests (Fig. 5) and the resulting environmental degradation, such as a decrease in species diversity, species extinction, erosion and in order to improve the nutrition of subsistence farmers, the W.E.I. initiated studies in 1976 towards the transformation, intensification and site-stabilization of shifting agriculture.

On the basis of these studies, project activities started in 1982 to combine traditional and improved appropriate gardening techniques and to extend these methods to officials, associations, village groups, and to the individuals who are expected to do training and extension-work.

The intended result of this program is to maintain an effective subsistence agriculture system which provides a self-sufficient food supply and supports the villagers socially, economically and nutritionally.

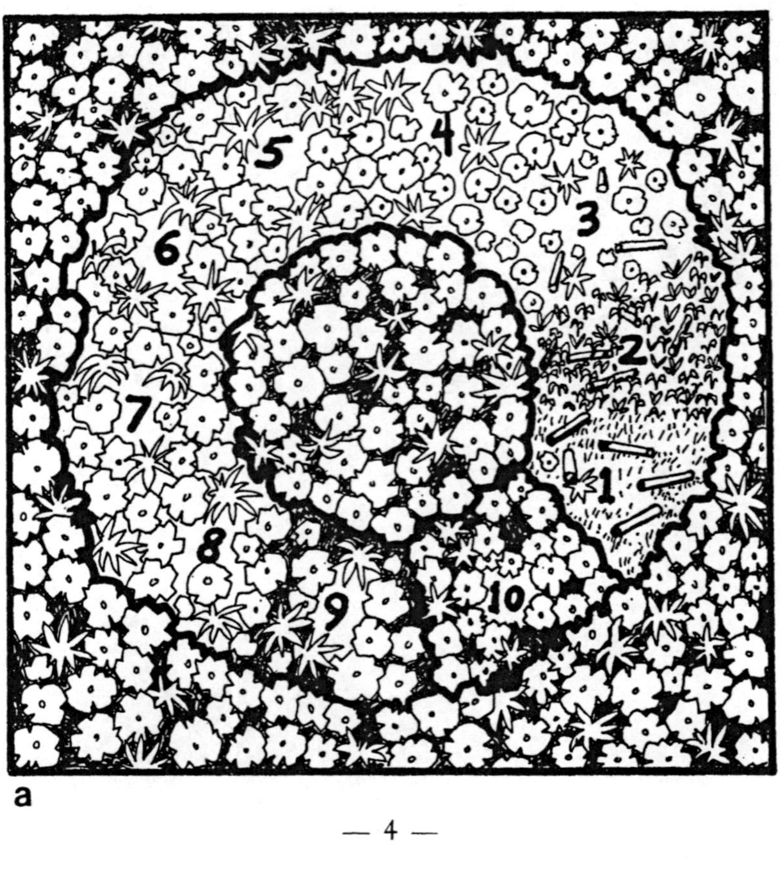

a

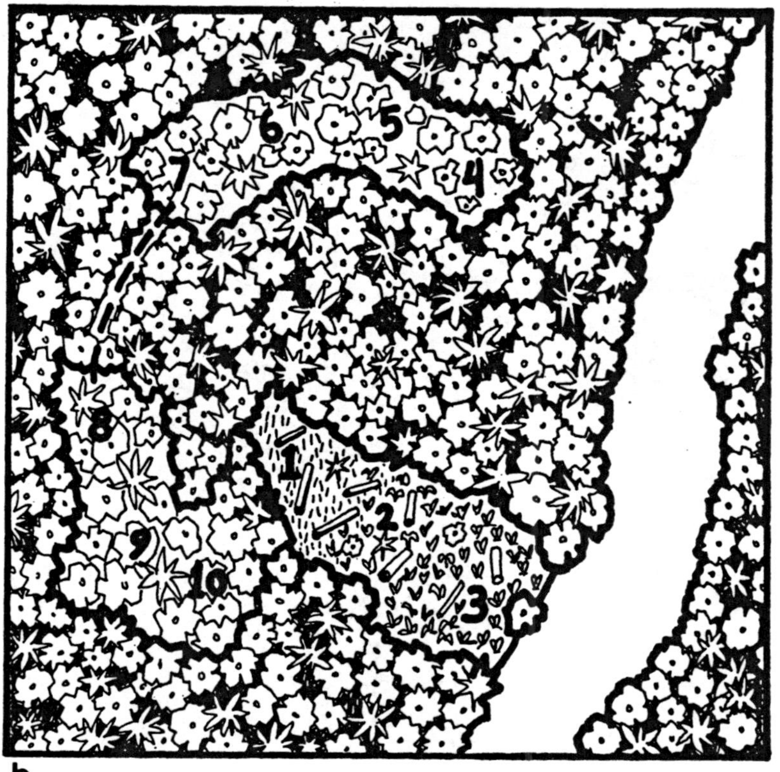

b

Fig. 3 Diagram of the two major bush-fallowing rotations used in shifting cultivation in Papua New Guinea.

a Principle of the rotation in circles around a village

b Random system of bush-fallowing due to accessibility from the riverside

1 Newly prepared area, 2-5 Garden area in use, 6-8 Regrowth of secondary forest, 9-10 Fully established secondary forest

The steadily increasing acceptance of the techniques developed by the Subsistence Agriculture Improvement Project of the Wau Ecology Institute lead to an official takeover of this project by the Provincial Government of the Morobe Province in Papua New Guinea.

Fig. 4 Subsistence systems in Papua New Guinea (PNG Atlas, 1978, changed) W Wau

Taro

Yam

Sweet Potatoe

Banana

Sago

Hunting

Fishing

Fig. 5 Cut and burn plot of a subsistence garden area in the Morobe Province in Papua New Guinea

Notes on subsistence agriculture and the subsistence agriculture improvement project

Traditional farming methods have evolved over thousands of years.

One traditional way of obtaining food is to gather the plants which grow wild (Fig. 6). But when people dig the land to grow a crop for their own food, this is called subsistence agriculture (Fig. 7). Subsistence farmers usually only grow enough food for their own family and all the food that is grown is usually used up. Subsistence farmers may not need to change their ways of growing crops, as long as enough land is available and the population density is low. A farmer may want to change if he needs more food for his family or the village people, or if he wants to have some more of his crop to sell.

If farmers come to a decision to sell crops they usually turn to cash crops like coffee, cacao, pyrethrum, cardamon or coconuts (Fig. 8). Then they can buy things they want. But this type

Fig. 6 Food gatherers

of agriculture is very different from subsistence agriculture and the farmer has to learn a lot about the soil, the plants and how they grow, how to keep pests and diseases away and how to get the crops ready for the market.

The cut and burn system or shifting cultivation system (Fig. 5) as it is practised now, gives the rural farmer very fertile soil virtually free of weeds. The crop can be grown without chemical fertilizers, dangerous pesticides and sophisticated tools. He will get a sufficient harvest if the rains do not fail.

Fig. 7 Subsistence farmers

This type of subsistence agriculture is a way of life in which people survive by growing local foodcrops in their gardens. Very simple methods and tools are used, in keeping with the natural environment. Traditionally, subsistence farming is done by a "cut and burn" method in regions with significant dry periods and by "cut and rot" method in very wet regions. Crops are usually sown without any specific soil preparation.

These systems are very suitable when there is a low population density, plenty of forest and many wild animals.

Because of changing conditions, the system has to change as well. This is why the Subsistence Agriculture Improvement Project supports many aspects of the traditional shifting cultivation system, but at the same time, changes it to a site-stable system.

To do so, the people involved in the project have always to bear in mind not only the practical handling of appropriately

Fig. 8 Modern cash cropping system

improved subsistence farming techniques, but the following principles:

— Go to the village and live with the people
— Search for the leading people in a village
— Love the people
— Learn from them
— Plan together with the people
— Serve the people

- Begin your extension work with what is already known by the villagers
- Build with local materials and tools

Each project co-worker and extensionist should keep in mind that "Our brothers and sisters on the bandwagon of the Subsistence Agriculture Improvement Project will change our attitudes towards rural folks from that of a superior public officer to that of a humble servant of people and commit and dedicate ourselves to do our part to eradicate the problems of Poverty, Hunger, Disease and Powerlessness".

The mid-mountain biogeography of Papua New Guinea

Topography and Geology

Papua New Guinea is the second largest island in the world. It is 800000 km^2 in area, 2400 km long, up to 720 km wide and the highest mountain reaches to almost 5000 m. A rugged mountain range forms the spine of the island (Fig. 9).

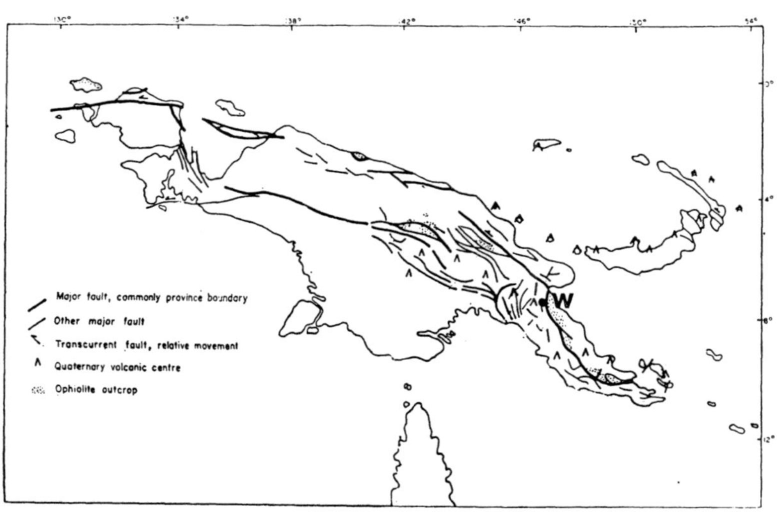

Fig. 9 Major structural elements of the Island of New Guinea. W Wau

Geologically, the country is young. The topography of the country is characterized by upgraded rivers, numerous waterfalls and rapids, stream piracy, steep valleys and sharp cliffs. The rugged terrain of the central range is at places interrupted by large elongated intermontane valleys. The mountain range falls steeply into the southern lowlands and alluvial and coastal plains form a composite delta. Northward, the mountains give way to wide east-west trending plains. The east Papuan Peninsula is mostly made up of the central range with the foot hills reaching the coast in many places.

The mid-montane area, where Wau is located is geologically called the Mobile Belt which consists of folded and metamorphosed oceanic sediments. Volcanic activity provided the main source of sediments containing gold, a characteristic of the Wau — Bulolo Area. Since 1888, about 166,000 kg of gold has been produced from this area (Fig. 10).

Soil

Soil formation from parent material is influenced by weathering, climate, vegetation and drainage.

Rankers and regosolic brown soils are predominant in the Wau Valley area. This association occurs on rugged mountains with steep slopes. Soil creeps are common and landslides occur. The soil profile is generally shallow. In addition, alluvial soils occur. These soils developed on recent alluvial deposits and have, if undisturbed, a high agricultural potential (Fig. 11).

In the lowlands most soils have a low natural fertility apart from the andosols and the well drained alluvial soils. The steep and unstable slopes in most of the mountain areas have a low potential for agriculture (Fig. 12). The average soil temperature at 50 cm depth between 1200 m and 2300 m is about 17°C. This is one of the lowest values when compared with other tropical montane areas of similar altitude (Dronia, 1983).

Climate

Papua New Guinea is a humid tropical island with moderate to very high rainfall. The environment is, in general, benign

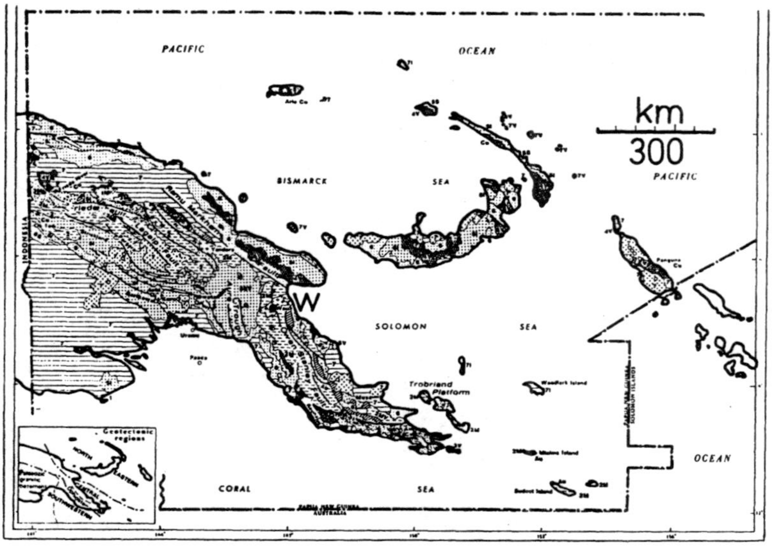

Fig. 10 Geology of Papua New Guinea
(PNG Atlas, 1978 changed) W Wau

7	Quarternary	—	Fault
6	Pliocene	▲▲	Thrust fault
5	Miocene	o	Gas discovery well
4	Paleogene	M	Metamorphic
3	Cretaceous-Eocene	V	Volcanic
2	Mesozoic	S	Sedimentary
1	Palaeozoic	I	Limestone
+	Granitic	Au	Gold
vv	Volcanic	Cu	Copper
	Ultramafic	Ni	Nickel
	Gabbroic	Mn	Manganese
		Cr	Chromium
		i	Intrusive

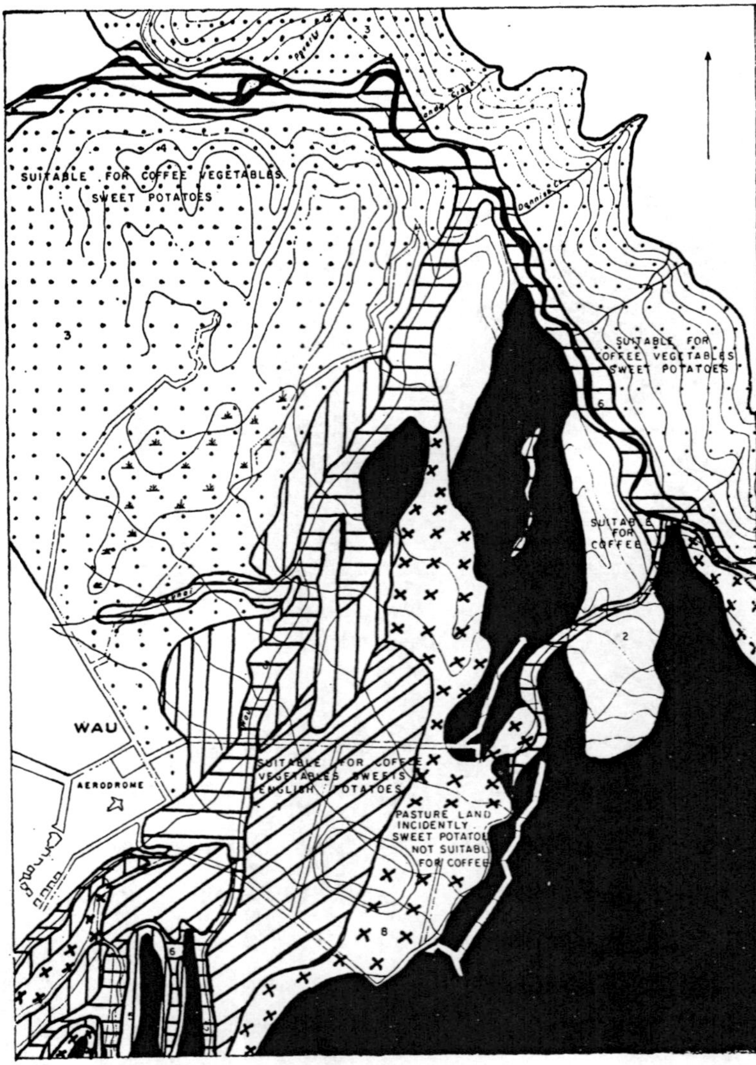

Fig. 11 Soil Map of the Wau Area

■ Soils developed from predominately lake bed clay
□ Soils developed from predominantly granite
⁚⁚ Soils developed from slate embeded in clay
▥ Soils derived from parent material consisting large quantities of granite, some slate and less clay
▤ Alluvial river sand
▨ Humic sand
☒ Alluvial clay

with moderate temperatures, high humidity, adequate rainfall most or all of the year, and light winds in many areas (Fig. 13).

The high cloud cover and an average solar radiation of 400 langleys per day is characteristic (Kalma, 1972). The climate is affected by equatorial westerlies, but surface winds are usually light (Bary, 1978).

Rainfall

It usually rains on both sides of the main cordillera of Papua New Guinea throughout much of the year. In general, there is less rain from May to August or September in the eastern mountains. Wau, at an average altitude of 1065 m above sea level, receives about 1700-1900 mm of rain per year. The distribution is shown in Fig. 14.

In Papua New Guinea, areas with a high rainfall (for example Lae), have a very wet season from June-September, the same time that mountain regions, like Wau, have their season of low rainfall (Fig. 15) (Mc Alpin et al., 1975).

The topography is of great importance, because winds are intercepted and deflected upwards by mountain ranges, where clouds drop most of their moisture. Therefore, the centers of inland valleys often have reduced rainfall (Fig. 16).

In many montane areas humidity is at the saturation point all night, so every night of the year is often foggy, as the daytime periods. It may dry to 50% or less on sunny days.

Temperature

The temperature seasonality is minimal, but the diurnal cycle is significant. The average temperature in Wau is 22.3°C including the absolute maximum temperature of 34.1°C and the absolute minimum temperature of 6.1°C. The average maximum temperature is 28.2°C and the average minimum temperature 16.5°C, showing the extent of the diurnal cycle (Fig. 17). The average temperature decreases by 0.6°C with each 100 m increase in altitude.

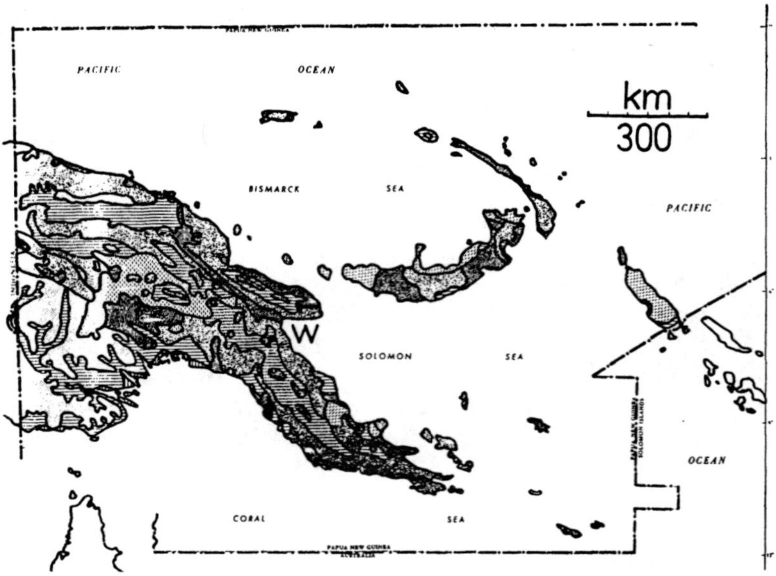

Fig. 12 Soils in Papua New Guinea (PNG Altas, 1978, changed)
W Wau

- Lithosols and alpine humus soil
- Rankers and regosols
- Rendzinas and Limestone soils
- Humic brown ash soils
- Latosols and brown soils
- Limestone soils
- Andosols
- Latosols and vertisols
- Well drained alluvial soils
- Poorly drained alluvial soils
- Mangrove soils

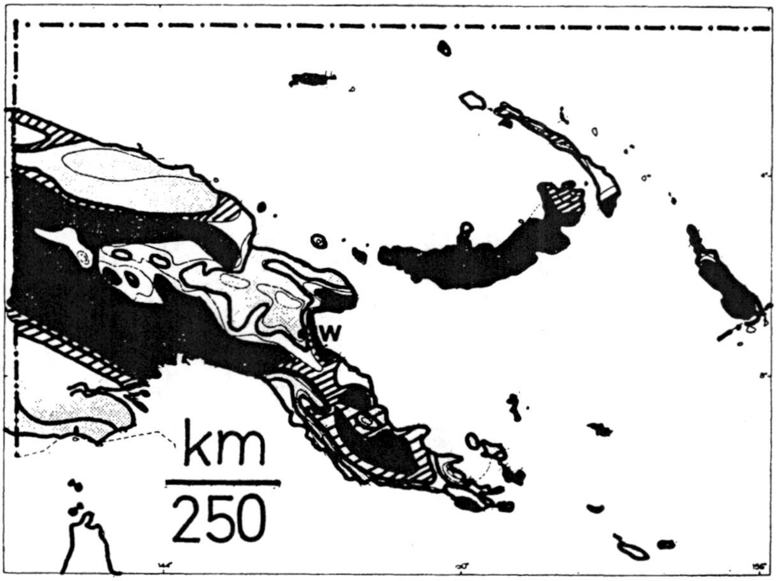

Fig. 13 Average annual rainfall in Papua New Guinea (PNG Atlas, 1978 changed) W Wau

- 3556+
- 2540–3555
- 1524–2559
- 0–1523

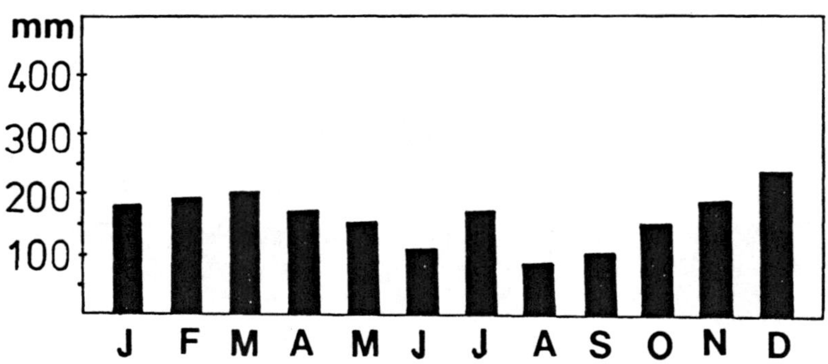

Fig. 14 Mean monthly rainfall in Wau, Papua New Guinea (average altitude: 1150 m; average total rain/year: 1843 mm; geographical location: 7°10'S, 146°44'E)

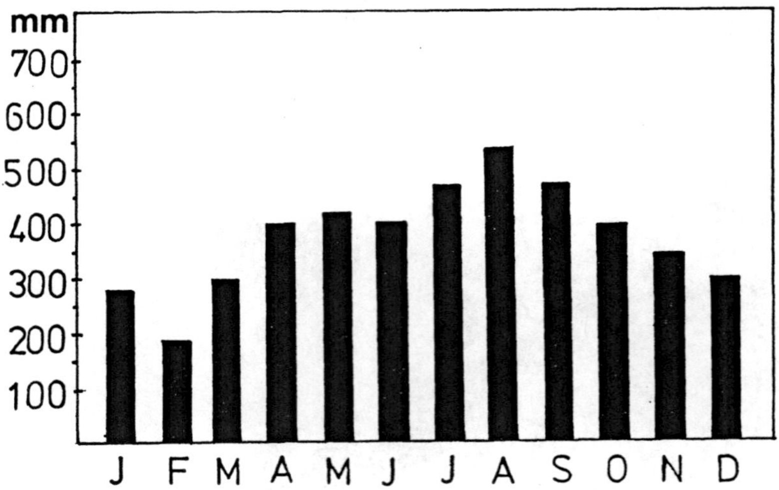

Fig. 15 Mean monthly rainfall in Lae, Papua New Guinea (average altitude: 10 m; average total rain/year: 4617 mm; geographical location: 6°44'S, 147°00'E)

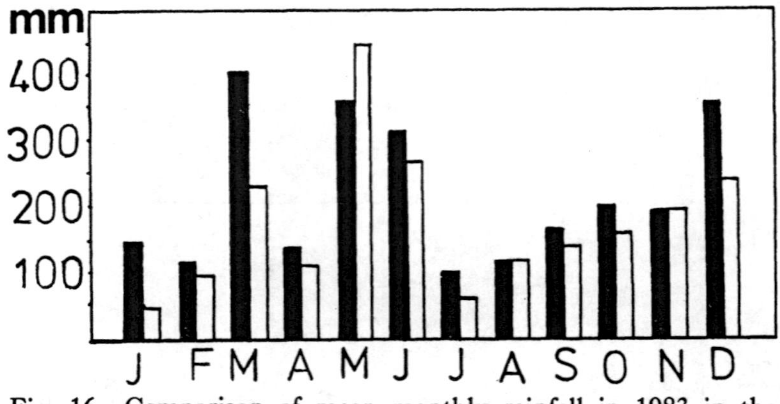

Fig. 16 Comparison of mean monthly rainfall in 1983 in the Wau Valley.

■ Station of the Wau Ecology Institute, located on the slopes of Mt. Kaindi at about 1200 m altitude

□ Station of the Katherin Lehman School, located in the centre of the Wau Valley at about 1000 m altitude

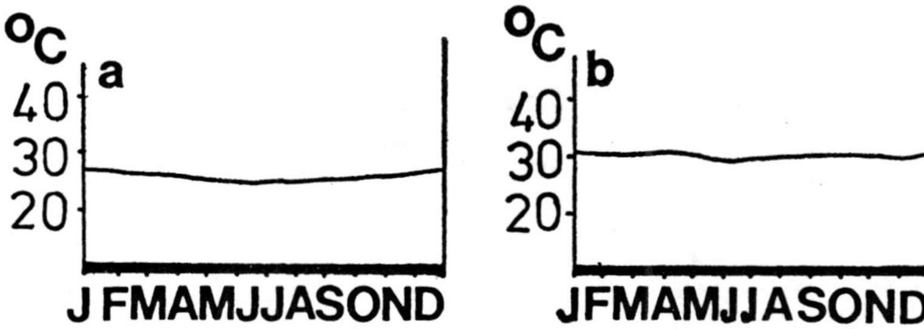

Fig. 17 Mean monthly temperatures in Wau (a) and Lae (b).

Appropriate tools for the subsistence garden

All tools and necessary basic equipment for the improvement of the subsistence garden are made from local material. As far as possible the local practises and the skills of the villagers and trainees were integrated in the program.

The construction of compost stockades

In constructing any compost stockade, three main points have to be considered:

1. A good compost heap should have a minimum size of little less than 1½ spade sizes in length, width and height, if arranged in a rectangular manner.
2. A compost heap must be protected against extreme weather conditions such as rainfall and direct sun.
3. It should be easy to get the compost in and out of the stockade.

The following four kinds of compost stockades will be described considering these three major points:

1. Walkabout stockade: For large scale production of compost.
2. Banis stockade: For small scale production of compost.
3. Wire stockade: For compost production in particular urban areas.
4. Plastic stockade: For compost production in sub-urban and urban areas.

The walkabout stockade

Fig. 18 The Walkabout stockade for compost production

1. Mark out a flat area of at least $1\frac{1}{2}$ spade sizes in width and $5\frac{1}{2}$ spade sizes in length. Make sure that on each side there is an area of at least 1 spade size left to build the roof and to provide enough working space (Fig. 19). A sloping area makes the shifting of the compost easier, by always moving it downslope. The construction of the stockade is, however, easier on a flat site.

2. Equalize the area with a spade.

3. Collect and prepare the following materials (Fig. 20):

   a. Spade or a pointed pole of about one pass in length.

   b. Bushknife.

   c. Wooden hammer or log to beat the sticks into the ground.

   d. Four posts of about $2\frac{1}{2}$ spade sizes in length and one fist thick, with a fork on the top and four similar posts of three spade sizes in length.

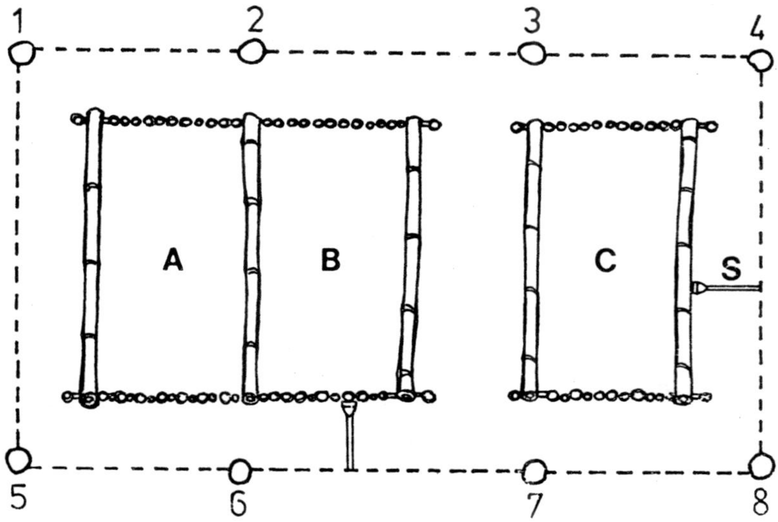

Fig. 19 Construction plan for the walkabout stockade
1 – 8 Supporting posts for the roof
A – C Stockade chambers
S Spade

e. Two straight poles about 6½ spades long and one fist thick.

f. Twenty-three sticks of 3 spade sizes in length and three fingers thick.

g. About 13 sticks of 6½ spade sizes in length. Flattened bamboo can be used.

h. Roof material like kunai grass, leaves of the pandanus or bamboo leaves.

i. About 60 bamboo poles or straight sticks about 1½ spade sizes in length.

j. About 90 poles, each about 1½ spade sizes in length plus one stretched hand (about 20 cm). Sharpen the posts with a bushknife.

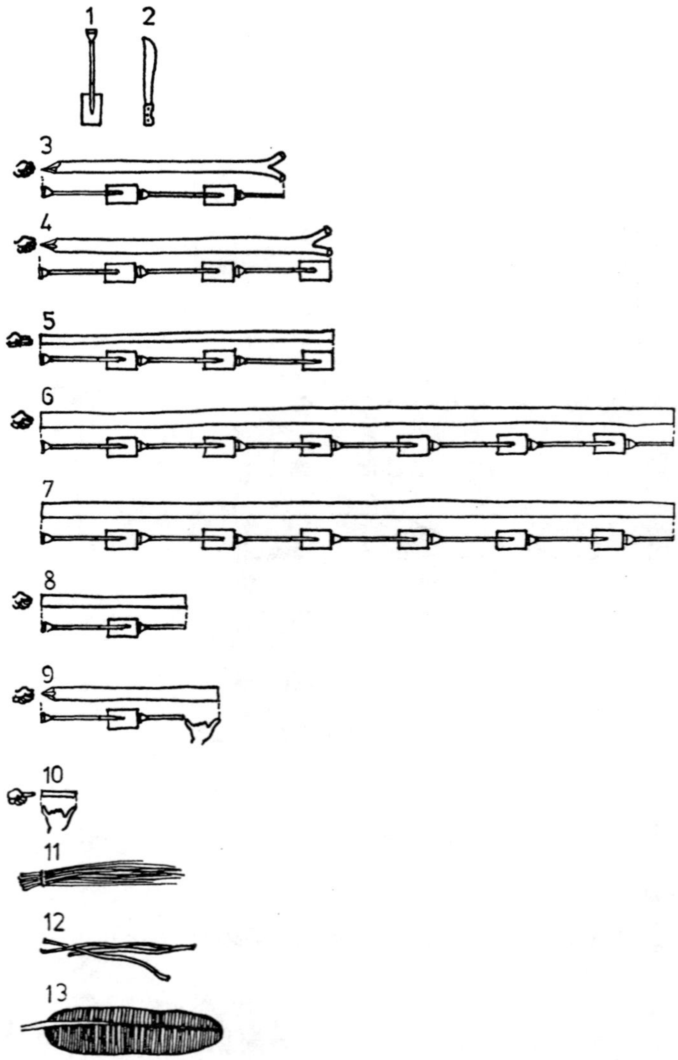

Fig. 20 Material needed for a walkabout stockade.
The numbers in brackets refer to the quantity needed. 1 Spade, 2 Bushknife, 3 Forked posts (4), 5 Straight poles (2), 6 Straight poles (2), 7 Straight poles or flattened bamboo (13), 8 Straight sticks or bamboo poles (62), 9 Straight sticks (90), 10 Straight sticks (80), 11 Kunai grass for the roof, 12 Ropes, 13 Banana leaves

k. About 80 sticks of approximately one stretched hand length and 1 finger thick.

l. Two poles or flattened bamboo sticks of 3 spade sizes in length and 2 poles with a length of 1½ spade.

m. Ropes, strings, bark or any other tying device.

4. The walls shown in figure 18 can be raised by hammering the 90 sharpened poles into the soil. Also shown in figure 18, a space large enough to easily insert a large bamboo pole has to be left open after pole 1, at the beginning of stockade two (about 1½ spade sizes in length from pole 1), at the beginning of the workspace area, and at the beginning and end of the third stockade.

5. To start the side walls, fit one bamboo of 1½ spade sizes in length between the first and the second pole.

6. Fasten one of the sticks above the bamboo pole with strings. Make sure that the opening under the stick is not narrower than the bamboo pole. The bamboo pole should easily be removable. Build the entire side-wall using this kind of construction until the top.

7. Build in the same way the other four side-walls of the stockade.

8. Fasten all parts as good as possible with strings, and make sure that the bamboo poles of all side-walls can easily be removed.

9. Construct the roof using the forked posts according to figure 19 and figure 20 and finish the roof by tying the twenty-three sticks over the entire length of the stockade (Fig. 21) and fix the roof material (e.g. kunai grass) to the flattened bamboo before fixing it to the roof.

10. The stockade is now ready for use. Further improvements can be made by placing some support poles as shown in figure 18 and 21 between the roof supporting poles. Make sure that they do not block the side-walls.

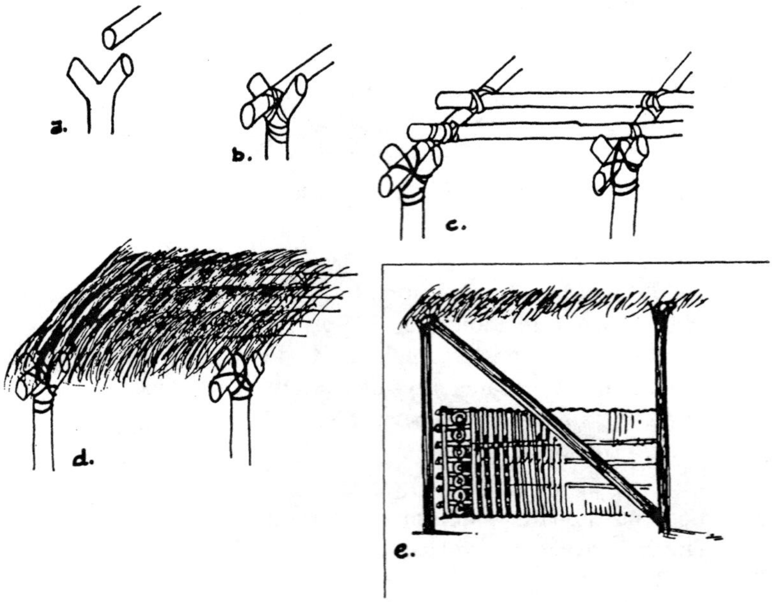

Fig. 21 Roof construction for the stockade and improvement of stability by supportive constructions

a–d Roof construction

e Setting of transversal part to improve the stability of the construction

This type of stockade may last for 5 years or longer with only minor repairs necessary.

The banis stockade

1. Put a stick in the middle of the flattened area where you want to set up the stockade. With a rope of approximately two stretched hands in length, mark the position of the fence by circling it around the stick in the center.

2. Collect and prepare the following material (Fig. 23):

   a. 30 wooden sticks of $1\frac{1}{2}$ spade sizes in length with sharpened ends.

Fig. 22 The Banis stockade for compost production

b. Flattened, bendable bamboo of about 3 spade sizes in length to construct the rims (at least 4 pieces).

c. 5 sticks about one stretched hand shorter than the 1½ spade sized sticks.

d. 2 forked posts for the top side of the roof of about 2½ spade sizes in length and 2 forked posts of 2 spade sizes in length, both with sharpened ends.

e. 4 sticks for the roof construction of about 1 spade size in length.

f. Roof material like kunai grass and bamboo sticks.

g. Ropes and strings from bark.

3. Hammer the wooden sticks into the ground along the circle marked. Leave an opening of about ½ a spade width.

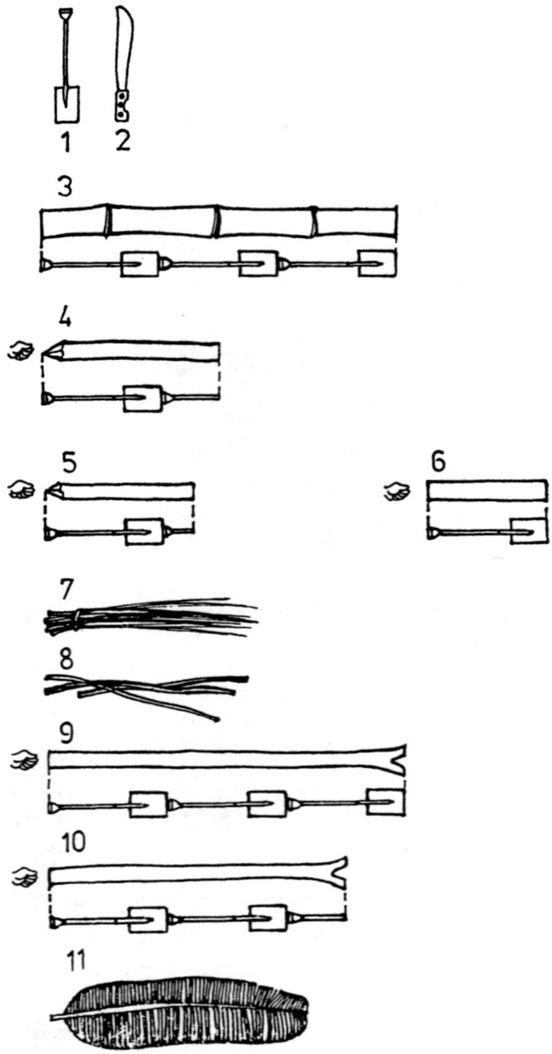

Fig. 23 Material needed for the banis stockade.
The numbers in bracket refer to the quantity needed. 1 Spade, 2 Bushknife, 3 Bendable flattened bamboo (4-6), 4 Pointed sticks (25-30), 5 Pointed sticks (4-5), 6 Straightsticks (4), 7 Kunai grass for the roof, 8 Ropes of bark, 9 Forked Posts (2), 10 Forked posts (2), 11 Banana leaves

4. Tie the flattened bamboo to the sticks and form an inner and outer rim round the top of the banis (Fig. 24). Any lower rim should leave open the opening.

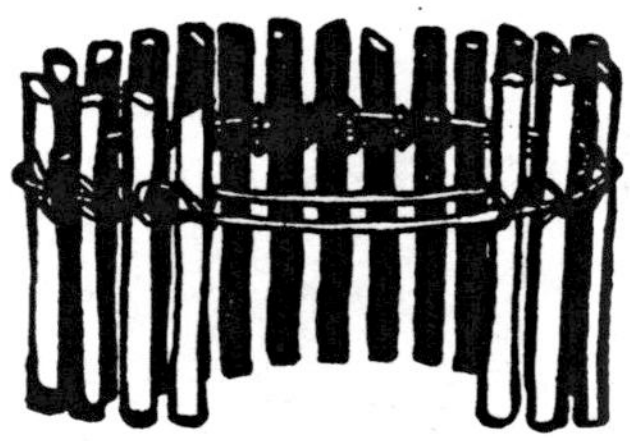

Fig. 24 Construction of a Banis stockade for the production of compost

5. Insert the removable shorter sticks in the opening.
6. Construct the roof so that the highside is above the opening. This makes it easier to move the compost in and out of the banis.

This type of stockade will last for 5 years or even more with only little repair necessary.

The wire stockade

Fig. 25 The Wire stockade for the production of compost

1. Any reasonable flat area can be used near the site you want to use the compost.
2. Collect and prepare the following materials (Fig. 26):

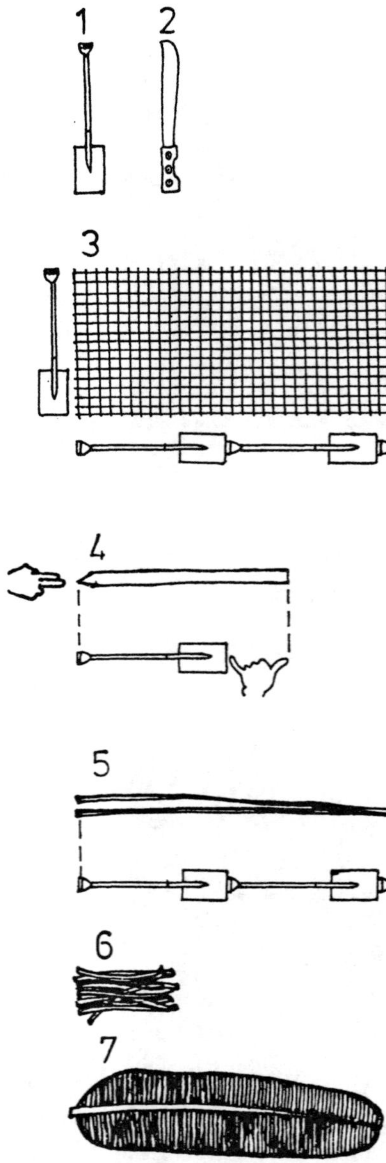

Fig. 26 Material necessary for the wire stockade.
The numbers in brackets refer to the quantity needed.
1 Spade, 2 Bushknife, 3 Sheet of mesh-wire, 4 Pointed sticks (7), 5 ropes, 6 Strings, 7 Banana leaves

a. Spade and bushknife.

b. A sheet of wire mesh of 1 spade size in width and 3 spade sizes in length.

c. Seven sticks of $1\frac{1}{2}$ spades sizes in length with sharpened ends.

d. Ropes and strings.

e. 15 large banana leaves.

3. Form a cylinder out of the wire mesh and let it overlap for about $\frac{1}{2}$ a spade length. Fix the parts together with string (Fig. 27).

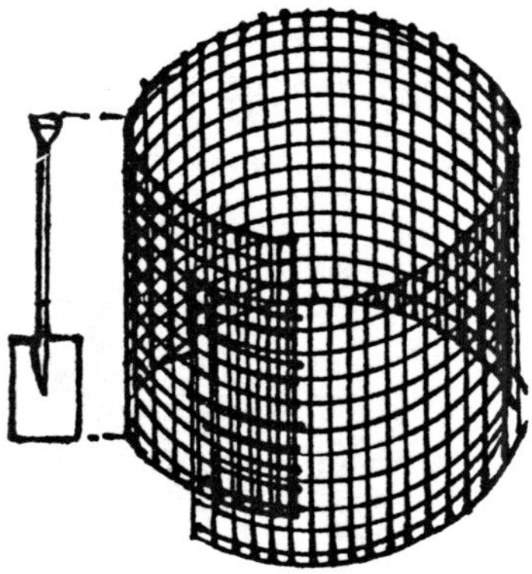

Fig. 27 Construction of the wire mesh stockade

4. Plant the 7 sticks inside the wire mesh (Fig. 28).

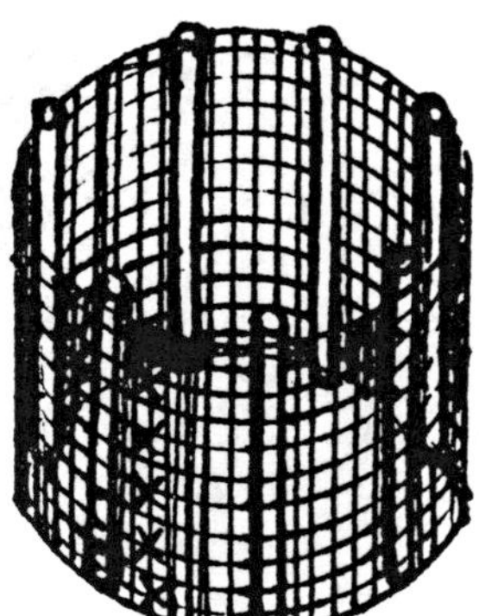

Fig. 28 Fixing of the sticks inside of the wire mesh stockade

5. Lift the wire mesh off and tie the sticks to each other with 3 rims of rope.

6. Place the wire mesh back and line the sides of the stockade with the banana leaves to form a parcel (Fig. 29).

Fig. 29 Placement of banana leaves inside of the wire mesh stockade

7. Place the compost material in the stockade.

This type of stockade can be used many times at any place the compost is necessary.

The plastic stockade

1. Any reasonable flat area near the site the compost is used can be used for this type of stockade.

2. Collect the following material (Fig. 31):

   a. Spade and bushknife.

   b. A black plastic sheet of about 6 spade sizes in length and two spade sizes in width or any other suitable size.

   c. 7 wooden sticks of 1½ spade sizes in length.

   d. Ropes.

Fig. 30 The plastic stockade for the production of compost

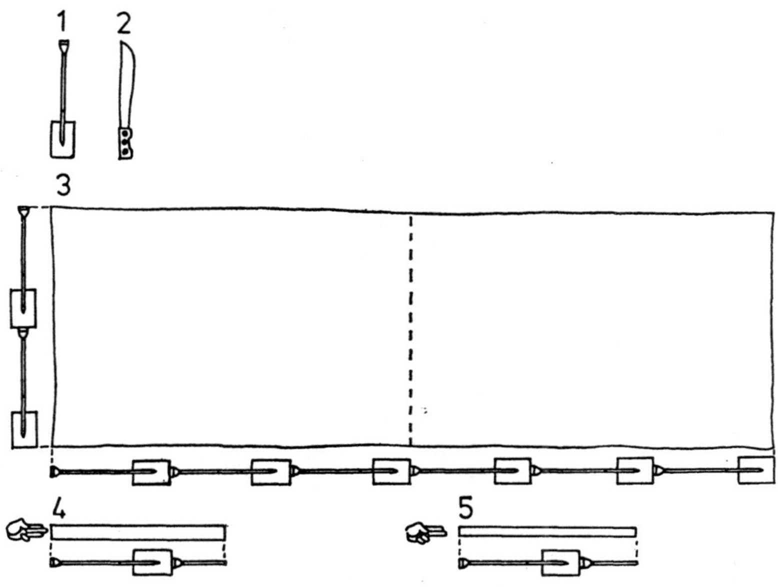

Fig. 31 Material necessary for the plastic stockade.
The numbers in brackets refer to the quantity needed.
1 Spade, 2 Bushknife, 3 Sheet of black plastic, 4 Sticks (6), 5 Sticks (7)

e. 5 Bamboo poles of $1\frac{1}{2}$ spade sizes in length and completely hollowed and holes cut at every knot (Fig. 32).

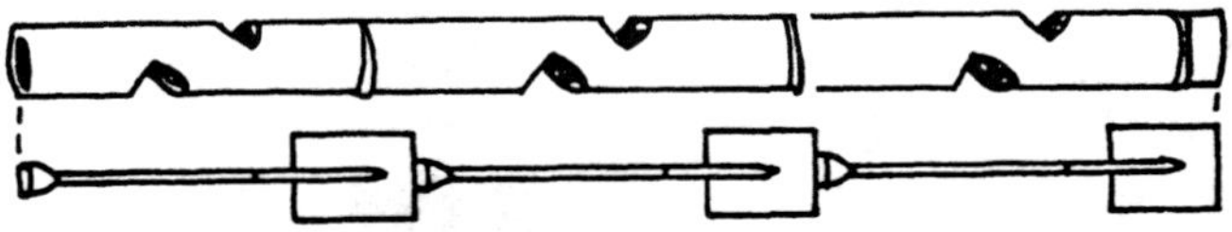

Fig. 32 Hollowed bamboo poles.
Five poles are necessary.

3. Hammer the 7 sticks at approximately equal distances in a circle into the ground.
4. Fix the sticks by combining them with two rims of rope, one near the bottom and the other near the top.
5. Cover the inside completely with the black plastic sheet and leave enough of the plastic sheet to cover the top of the compost heap after being piled up.
6. While filling the stockade with compost, insert the hollowed bamboo poles necessary to aerate the compost during the rotting process (Fig. 33).

Fig. 33 Plastic stockade with inserted hollowed bamboo poles ready to be closed.

7. Tie the top of the plastic to cover the compost completely.

Construction of a nursery

There are many different ways possible to construct a nursery. The main requirements for any nursery are the following:

— Protection from too much sun and rain.
— Good soil mixture necessary: fertile, deep enough, not too sticky.
— Availability of water near the nursery site.
— Protection from weeds, pests, diseases.
— Protection from animal attacks.
— Size adjusted to workability: e.g. not too wide beds.

Nursery for vegetables

Fig. 34 Nursery for vegetables

A nursery with seedbeds for vegetables, which usually spend only 4 – 8 weeks in the nursery, can be constructed as follows:

1. Decide on, how many plants the nursery should be able to contain at one time. For example:

    — You want to grow tomatoes.
    — You have an area of 10 spade sizes in width and 10 spade sizes in length.

- The distance between the tomato plant rows should be 1 spade size in length, and the distance between the single plants should be about ½ a spade size in length
- In one row of 10 spade sizes in length you can grow $2 \times 10 = 20$ plants – 1 plant = 19 plants.
- Ten rows are available, therefore you have to multiply 19 plants times 10 = 190 plants. This is the necessary number of plants to be grown in the nursery.
- As you know not all seeds will produce seedlings. About ¼ of the seeds will not germinate. Therefore you need to plant in your nursery 250 seeds to get 190 seedlings. It is better to sow at least 270 seeds because you then can replace the plants which die due to unexpected reasons.

2. Calculate the approximate area you will need for your seedlings in your nursery. For example:

   - The distance between seeds is usually a thumb length (about 5 cm).
   - For 270 seeds you need therefore about 1½ spade sizes in length and ½ a spade in width to sow the seeds in 10 rows, parallel to each other.

This example will give you an idea how big your nursery should be to serve all your purposes. The following construction plan is based on a reasonable sized nursery for multiple purpose germinating of vegetables for a medium sized garden.

3. Mark out an area of about 8 spade sizes in length and 1½ spade sizes in width.

   The area should be located in a central place in your garden, preferably close to a permanent water supply. The area and surroundings must be carefully cleaned and cleared of weeds and possible host plants for pests and diseases. If animals are in the vicinity, a fence is necessary leaving at least 2 spade sizes as working space around the beds.

4. Collect and prepare the following material (Fig. 35):

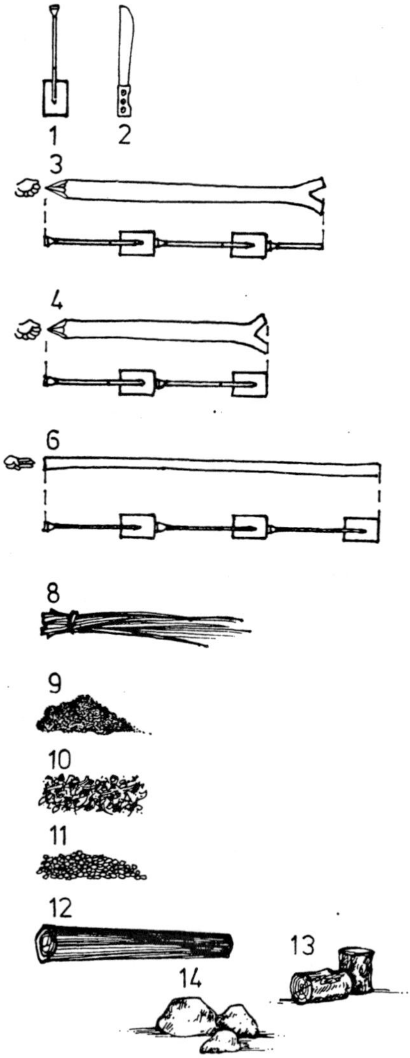

Fig. 35 Material necessary for the construction of a vegetable nursery.
The numbers in brackets refer to the quantity needed.
1 Spade, 2 Bushknife, 3 Forked poles (4), 4 Forked poles (4), 5 Straight posts (2), 6 Straight sticks (30), 7 Flattened bamboo (13), 8 Kunai grass, 9 Sterilized soil, 10 Sterilized compost, 11 Sand, 12 Banana trunk, 13 Logs, 14 Stones

a. Bushknife and spade.

b. Four strong, forked poles of 2½ spade sizes in length and four strong, forked poles of 2 spade sizes in length as roof supporting posts. They should be sharpened at the end.

c. Two straight posts of about 8 spade sizes in length and about one fist thick.

d. About 30 sticks of 3 spade sizes in length and about 3 fingers thick.

e. About 13 flattened bamboo pieces of 8 spade sizes in length.

f. Roof covering material like kunai grass.

   If the roof is used to collect water, place the corrugated iron sheets on top of the kunai grass layer, which will serve as insulating material.

g. Pre-sterilized soil, mixed with sterilized compost and sand.

   If no sterilized soil is available, leave the soil for a couple of days in the sun to kill unwanted germs and seeds.

h. Logs and banana trunks or stones to line the total area of the nursery.

5. Place the ready prepared soil for the nursery in the marked area and line it with logs, banana trunk halves or stones.

6. Build a roof but make sure that there is at least a ½ spade size left as working space around the bed, but protected by the roof.

7. For every load of seeds to be planted, separate beds should be prepared always using a fresh layer of topsoil (compost mixed with soil). Separate the various beds with halved banana trunks fixed to the ground with sticks (Fig. 36).

Fig. 36 System of separation of various seedbeds in a nursery

8. Cut the prepared and pre-watered bed into bricks, preferably of compost, with the bushknife. Each brick should have the size of 5 × 5 cm and one thumb length on each side. Plant 3-4 seeds in the centre. Later two plants should be removed. This kind of set up is recommended if the seedling will be transplanted into poor soil or if seedlings are produced for sale.

The use of banana-fibre-pots (see Fig. 42) in the nursery makes the control and transport even easier.

Nursery for trees

Fig. 37 Nursery for trees

For tree seedlings a good friable soil is even more important, because these seedlings stay for a much longer period in the nursery. Usually it is necessary for them to stay for several months (e.g. some fruit bearing trees) or several years (e.g. for reforestration purposes or coffee).

To be able to adjust the nursery according to the requirements of the seedlings during their time there, the roof must be constructed in a way that can easily be re-arranged according to the plants' growth patterns and working requirements in the nursery.

Further, each single plant should be grown in a separate pot, preferably made from banana fibres (see Fig. 42), which have the advantage of being homemade, cheap and will rot after being transplanted, serving the plant as a sort of compost without disturbing the root development. If the plant is growing bigger and bigger, only bigger and bigger pots have to be used, planting the old pot into the new one. The only disadvantage is that they will rot sometime before the young trees are ready to be transplanted, but a bigger pot is then the only additional necessity.

For a nursery housing about 300 trees seedlings the following is necessary:

1. Collect and prepare the following material (Fig. 38):

    a. Bushknife and spade.

    b. Six forked strong poles of about 2 spade sizes in length.

    c. Two poles of 3 spade sizes in length.

    d. 3 sticks of 1 spade size in length.

    e. Flattened bamboo of 3 spade sizes in length to cover the nursery.

2. Equalize an area of 3 spade sizes in length and 1 spade size in width. Make sure that no surface water can enter the area by digging a drainage canal around the nursery.

3. Dig appropriate holes for the 6 poles around the area and plant the posts as good as possible in position.

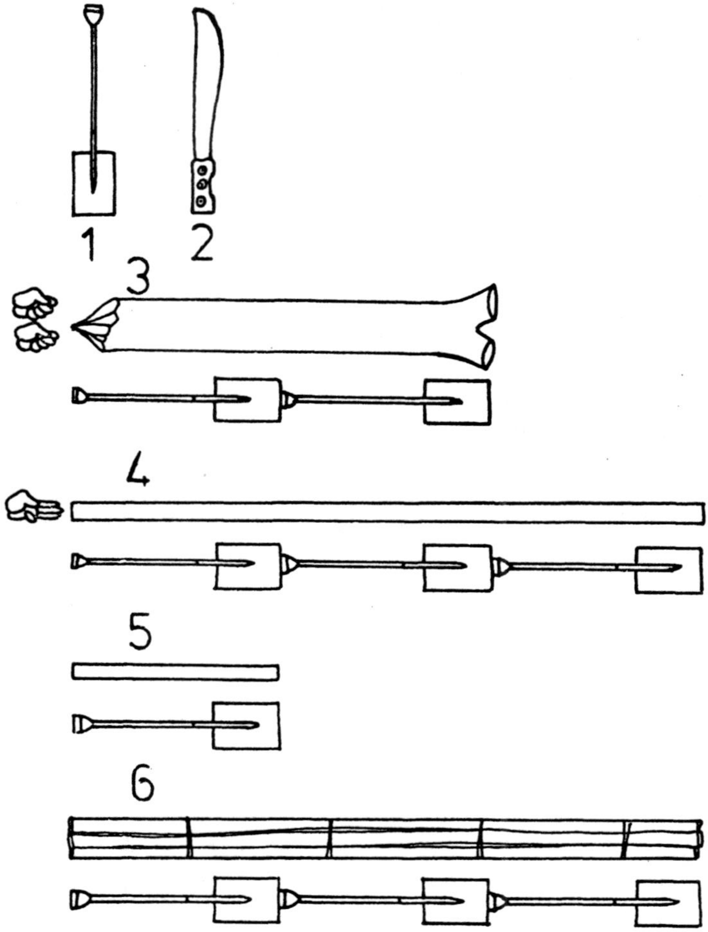

Fig. 38 Material necessary for the construction of a tree-nursery. The numbers in brackets refer to the quantity needed. 1 Spade, 2 Bushknife, 3 Forked Posts (6), 4 Straight posts (3), 5 Sticks (3), 6 Flattened bamboo

4. Fix the long posts to the supporting poles (Fig. 39) and cover the construction with the flattened bamboo and fasten it very carefully to the transverse sticks. If the plants become bigger and taller, make sure you can remove part of the flattened bamboo, because the plants need to adapt themselves to field conditions.

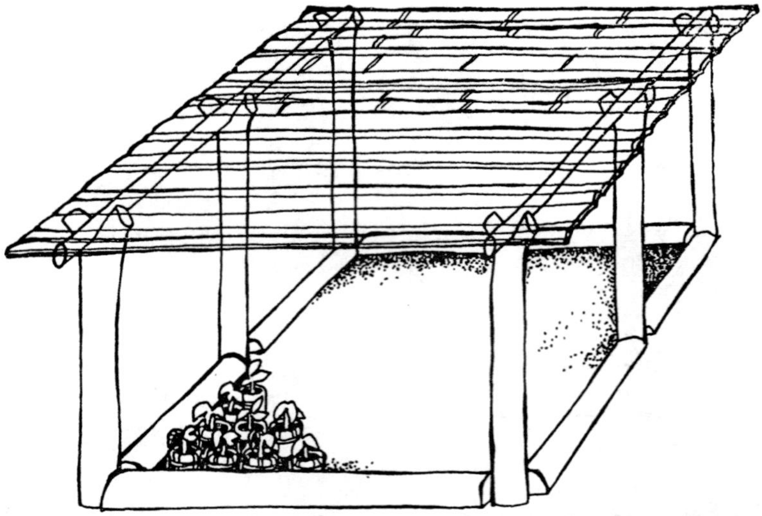

Fig. 39 Final set up of the nursery. The soil is covered with sawdust to suppress weed growth.

5. Plant the seeds in the soil/compost mixture in the banana-fibre-pots and line the area of the lined up pots with tree trunks or stones. Leave a walking space between the pots as convenient.

Preparation of compost bricks

Before starting to make compost bricks you should decide, whether the additional efforts and labour involved is reasonable or not. For vegetables such as tomatoes, capsicum, lettuce and cabbage etc, which all have small seeds and a relatively low germination rate, compost bricks can be recommended particularly, if you want to produce the seedlings for sale. For all other crops it is enough to use composted soil only. If compost bricks are the right choice then follow the following procedure:

1. Produce compost with a very fine texture and sieve it. About one banis stockade of compost is necessary for about 500 bricks.

2. Dry the compost thoroughly, preferably in a solar drier (see Fig. 56).

3. Collect black soil and sieve it; also sieve sand or fine gravel.
4. Sterilize the mixture of soil, compost and sand in a drum sterilizer (see Fig. 46). Mixture ratio: 1 spade of soil + 1 spade of compost + $\frac{1}{4}$ spade of sand.
5. Collect and prepare the following material (Fig. 40):

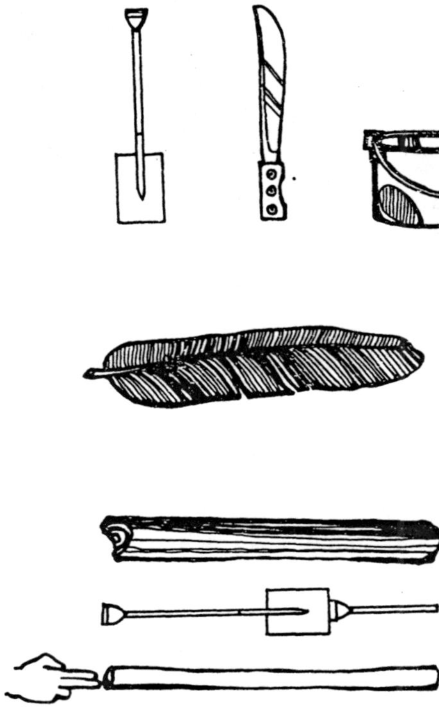

Fig. 40 Material necessary to produce compost bricks.
The numbers in brackets refer to the quantity needed.
1 Spade, 2 Bushknife, 3 Waterpot, 4 Banana leaves (4 – 6), 5 Halfed banana trunks (4), 6 Wooden liner (1)

a. Bushknife and buckets with water.

b. Banana leaves and some slitted banana trunk halves of $\frac{1}{2}$ a spade size in length.

c. A piece of cardboard of one finger or one thumb in width and $\frac{1}{2}$ a spade size in length. This is necessary to equalize the mixture.

6. Prepare a rectangular area in the nursery by lining it with the halved banana trunks. Fix the trunks with sticks to the ground. Cover the bottom of the area with banana leaves (Fig. 41).

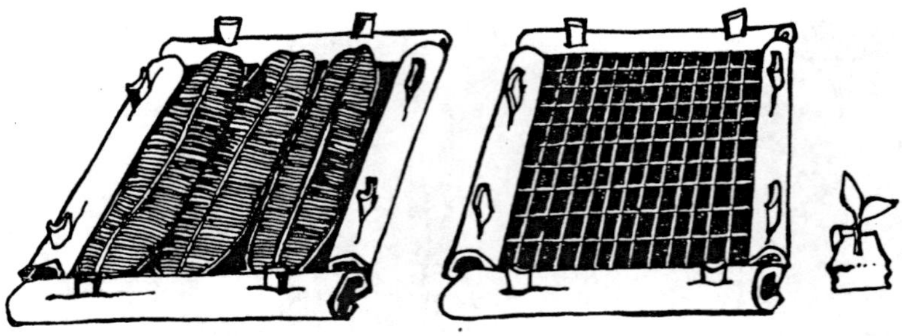

Fig. 41 Set up for the preparation of the compost bricks in a nursery

7. Mix the sterilized compost brick-mixture with water to form a paste. Pour the mixture into the area in the nursery, equalize it and cover it with banana leaves and let it dry for about two days.

8. If the cutline of the bushknife stays clear, the mixture is ready to be cut into bricks of one thumb width and one thumb length (5 cm × 5 cm). Use the piece of hardboard to do the cutting. Dip the bushknife after each cut into water.

9. Before placing 2–6 seeds per brick, let the bricks dry for about 2–3 hours.

10. Daily watering of the bricks is necessary by sprinkling with a watering can.

Set up for the production of banana-fibre-pots

The best place to set up a banana-fibre-pot site is in the shadow of a tree near your house or the compost stockade. There you should construct a bench and hammer at least four posts of different, diameters into the ground so that you can reach them comfortably if you are seated on the bench (see Fig. 42). Your knees should then be left and right of the post.

Fig. 42 Set up for the production of banana-fibre-pots

Then collect the following materials (Fig. 43):

1. Flattened and dried skins of banana trunks. Drying time: 1–2 days

— Ropes from Hibiscus bark or other kinds of lasting ropes.

To produce the banana-fibre-pots, follow the steps given in Figure 45:

1. Soak the dried banana fibres in water until they are soft again.

2. Decide how tall your pots should be and mark this with a cut at the respective post (Fig. 44).

3. Cut the fibres four times as long as the desired height of the pot.

4. Place the skins over the stick as shown in Figure 45.

5. Tie the crosswise placed fibres about halfway along the length.

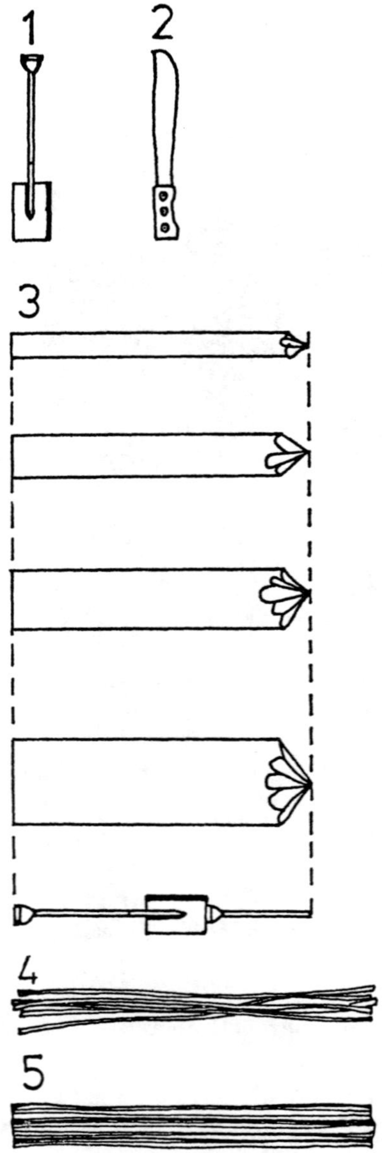

Fig. 43 Material necessary to produce banana-fibre-pots
1 Spade, 2 Bushknife, 3 Variety of different sized posts,
4 Ropes, 5 Flattened and dried banana-trunk-fibres

Fig. 44 System of marking the banana-fibre-pot size

Fig. 45 Procedure for the banana-fibre-pot production

6. Fold the remaining parts upward but make sure that the fibres are not moved upwards or in any other direction. The bottom of the pot should be flat.

7. Tie one string about one finger from the bottom end of the pot and another string one finger from the upper rim of the pot.

8. Cut the fibre ends according to size and remove the pot gently.

9. During storages of the pots, fill them with leaves or soil to hold them in shape and keep them in a shaded place.

The construction and function of a soil-steamer-drum

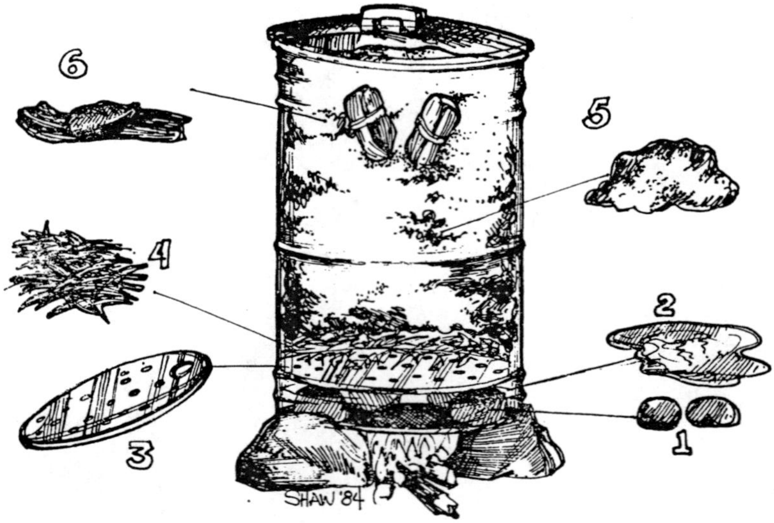

Fig. 46 Drum-steamer to sterilize soil and compost
1 Stones, 2 Water, 3 Drum-lid with holes, 4 Grass, 5 Soil or compost, 6 Sweet potato wrapped in banana leaves

1. Collect the materials shown in figure 46 and a strong old bushknife or metal scissors.

2. Cut the lid of the drum and make small holes in it. Then flatten the sharp edges of the rim of the drum and on the lid.

3. Place the drum on top of some stones so that you can place a fire under the drum.

4. Put some smaller stones to support the lid-raster.

5. Pour water into the drum so that the lid-raster is not under water and cover the raster with grass to prevent the soil or compost dripping into the water.

6. Fill the drum with the material which you want to be sterilized and place some sweet potatoes wrapped in banana leaves on top.

7. Cover the top with banana leaves and start the fire underneath the drum.

8. After about 2 hours you should check whether the sweet potatoes are ready cooked. This indicates whether the soil is already well steam-sterilized or not.

9. Let the sterilized material cool down before being used and keep it protected from rain, insects and weed seeds in a well covered bucket or drum.

Construction of an appropriate water-can-buffle

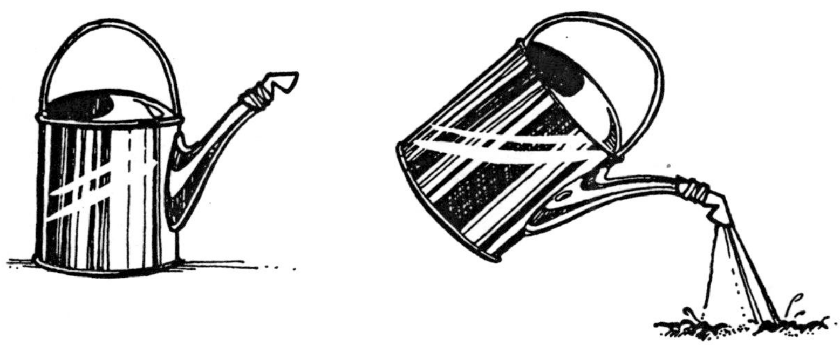

Fig. 47 Appropriate water-can-buffle

Very often the buffles of commercial water cans get stuck with dirt or they break, allowing no further sprinkling of water. To replace them, one can construct a buffle from any old tin lid.

1. An old round tin lid, a knife and a rope is necessary for the construction.

2. Make two incisions in the cut out lid as shown in Figure 48.

3. Fold the prepared lid in that way that the waterbeam from the sprout is dispersed as required. Fix the lid-buffle with the rope tightly to the sprout of the watercan.

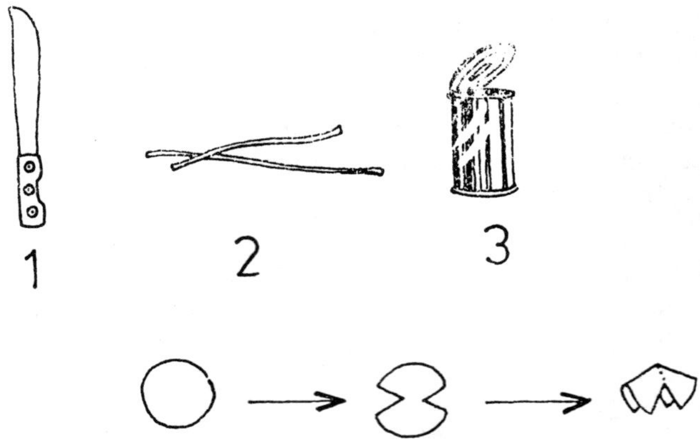

Fig. 48 Construction of a tin-lid buffle and material necessary
1 Bushknife, 2 Ropes, 3 Old round tin

Construction and use of a bamboo-marker

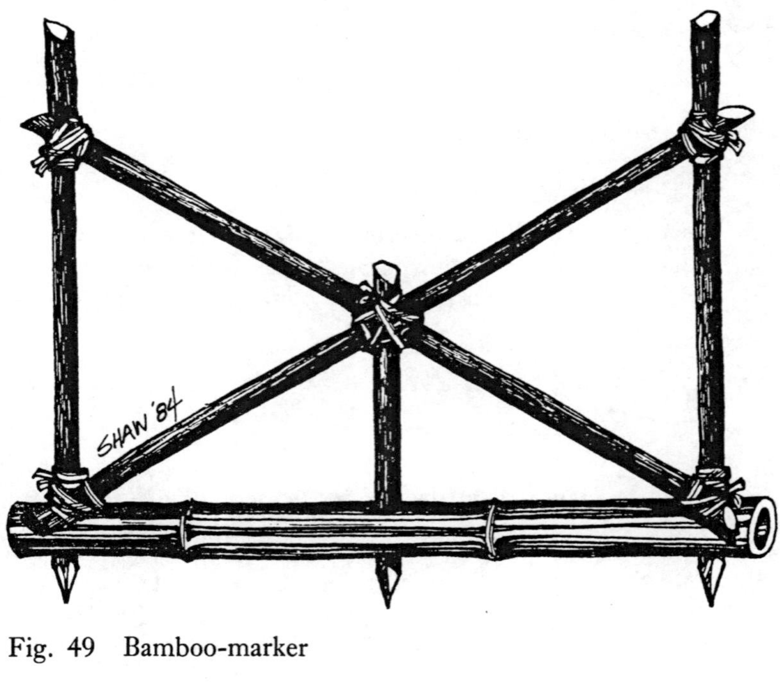

Fig. 49 Bamboo-marker

1. Collect and prepare the following materials (Fig. 50):

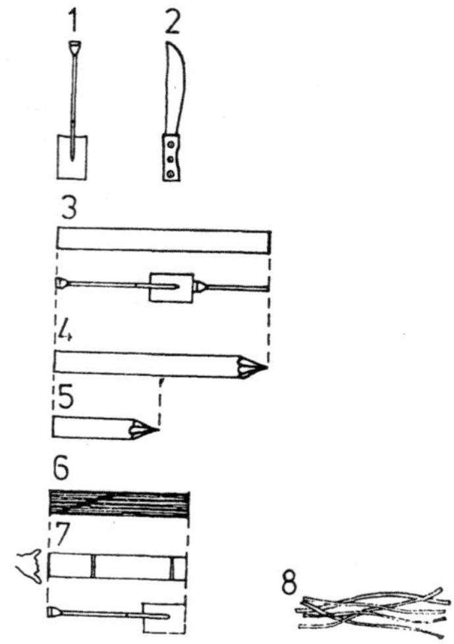

Fig. 50 Material necessary to construct a bamboo marker
The numbers in brackets refer to the quantity needed.
1 Spade, 2 Bushknife, 3 Straight stick (2), 4 Pointed sticks (2), 5 Pointed stick (1), 6 Banana trunk or 7 bamboo, 8 Ropes

a. Bushknife and ropes

b. Piece of bamboo about 1 spade size in length and 1 fist thick (a banana trunk may also be used)

c. Two strong, straight sticks of $1\frac{1}{2}$ spade sizes in length

d. One shorter stick of $\frac{1}{2}$ a spade size in length

e. Two sticks of $1\frac{1}{2}$ spade sizes in length but sharp pointed

2. Sharpen the sticks and place them through the bamboo or banana trunk according to the distance you want to plant your seedlings in the field or to sow your seeds in the nursery.

3. Fit the other sticks crosswise between the standing sticks to support the structure.

4. To mark the positions of the plant rows you pull the marker towards yourself as shown in Figure 51.

Fig. 51 Use of marker

If you cultivate in beds, you should mark the rows perpendicular to the bed to facilitate weeding (Fig. 52).

Make sure that the space between the plant rows is large enough to walk through the mature crop without damaging it. It is recommended to place the rows in bigger fields either along the contour or in the direction of the greater length to facilitate weeding and harvesting.

The A-frame construction and its use

One of the major problems in gardens, often placed in sloping areas, is soil erosion caused by water run-off. One of

Fig. 52 Use of marker in a garden bed

the most effective counter measures is to construct the gardenbeds in leveled terraces or in contour ridges or to construct them along the contour of the area. For all of these three methods it is necessary to find those points in a given area which have exactly the same altitude. By combining these points and constructing the gardenbeds along these contour lines, the detrimental effects of soil erosion can be greatly reduced.

1. Collect and prepare the following materials (see Fig. 53):

   a. Bushknife and ropes.

   b. Two sticks as long as a man with his arm above his head (see Fig. 53). Mark these sticks at the belly-nipple height, with the bushknife.

   c. One stick as long as the distance from the tip of a stretched hand and arm to the shoulder of the other arm (see Fig. 53).

   d. One small stone tied to a rope.

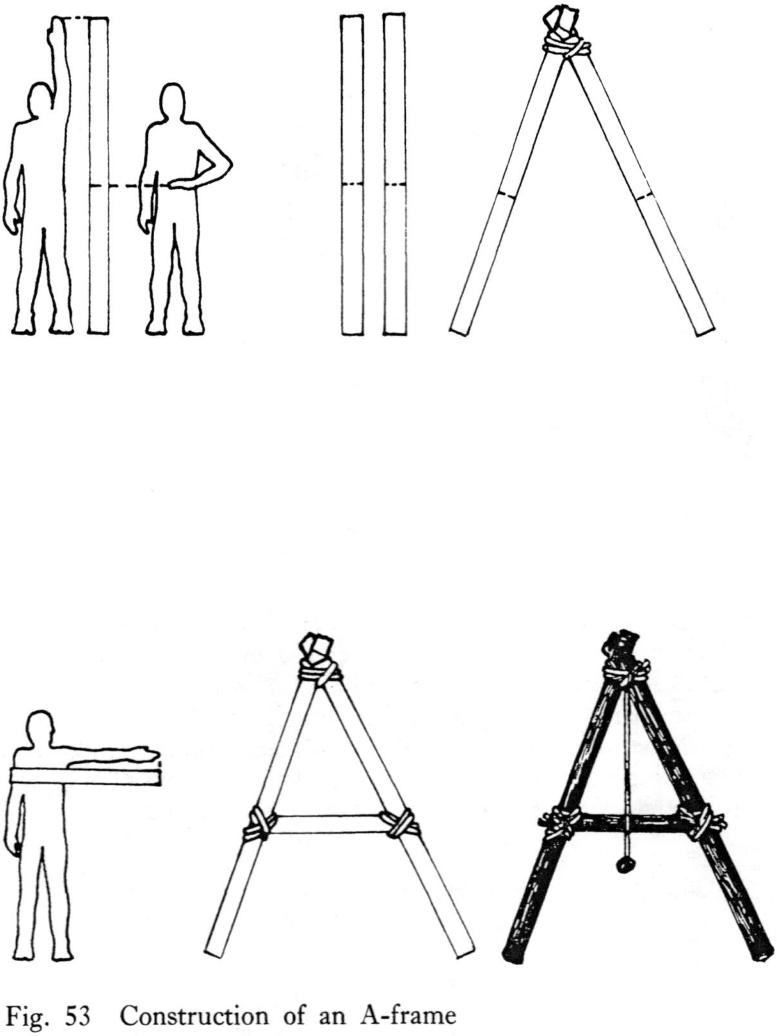

Fig. 53 Construction of an A-frame

2. Tie the two long sticks together as shown in Figure 53, but do not yet fasten the rope too tightly.

3. Cut out a half oval at each end of the short stick about three fingers width from each end.

4. Tie the short stick with its prepared ends to the longer sticks at the marks at belly-nipple height. Fasten the top-rope of the A-frame.

5. Find the exact middle of the horizontal stick by measuring the total length with a rope, half the rope by folding it back and mark the found middle with the bushknife (fig. 54).

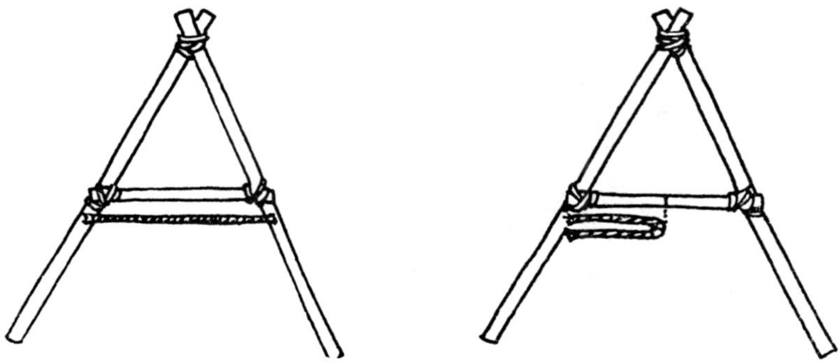

Fig. 54 System to find the exact middle of the A-frame

6. Tie the rope with the fixed stone in that way to the top of the A-frame so that the stone hangs about one fist from the horizontal line.

7. Collect some sticks and sharpen their ends before you start the A-frame in the selected site for your terrace or contour ridge.

8. Place one of these sticks into the ground where you want to start the first garden bed or terrace.

9. Put one leg of the A-frame close to the stick (Fig. 55) and move the other one until the stone is exactly in the middle of the horizontal stick. Place another stick into the ground at the place found to be at the same level as stick No. 1.

   Now repeat the procedure with the second stick and continue until you have reached the end of the terrace or contour ridge.

10. Combine the different sticks. By this means you will get the base line for your terrace or contour ridge No. 1. Now move further up-hill and measure the next one in the same way and so forth.

Fig. 55 Use of A-frame

A simple solar drier

As shown in Figure 56, a solar drier for seeds can be constructed by building a wooden frame and covering it with white plastic. The racks in the frame should be constructed out of flattened bamboo so that the hot air can easily move through the racks. The seeds should be placed on top of newspaper on the racks. Do not use plastic sheets. For more details, refer to the information published by:

SPATF,
P. O. Box 6937,
Boroko,
Papua New Guinea.

Fig. 56 Simple solar drier made from bamboo with a cover of white plastic

Seed selection and storage

This is the description of the methods of how to care for the seeds coming from your garden.

There are a variety of methods to get the seeds and ways to store them. Here we will describe one possible way.

1. Selection of seeds

Check the conditions of your plants in your garden by observing the following:

— Does the plant produce high or low yields
— Are the fruits big or small
— Is the amount of fruit large or small
— Is the plant growing well or not

If you check your plants this way you are able to select the plants with high yields, big fruits and good growth from their seeds.

Because seeds, like plants, can get diseases and can be damaged by insects or by various other pests, it is necessary to handle the seeds from your selected plants properly.

2. Drying and storage of seeds

The selected seeds must be carefully cleaned and dried. Drying is always necessary if you want to store the seeds for a long period. Therefore, the seeds selected must be dried in a solar drier where no rain can interrupt the drying process (see Fig. 56).

Wet seeds will rot as with all organic matter when wet, or else they will germinate and subsequently die.

Another way to dry the seeds is to place them on a metal sheet and expose them to the sun. Make sure that no animals eat the seeds.

A good traditional way to prevent seeds from being destroyed is to smoke them. By this way insect attack is reduced.

The dried seeds should be stored in a tin or plastic bag which is absolutely waterproof. The storage place itself should be dry and out of the reach of animals such as chickens.

Terrace and ridge construction

With the help of the A-frame, mark the contour-lines of your garden area where you want to set up the terraces or ridges.

Terrace construction

Build a wall of stones or logs along the marked line and fill the space upslope of the wall with the soil, dug from the upslope end of the terrace to be (Fig. 57). Let the terrace slope slightly towards the mountainside for proper drainage without causing soil erosion. Mark another contour line 2 spades up or down the slope and repeat the procedure until the whole garden area is ready for planting (Fig. 58) and (Fig. 63).

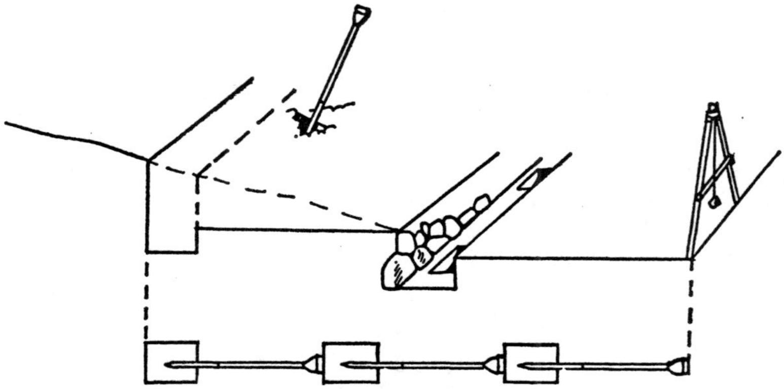

Fig. 57 Construction of a terrace with side walks and silt traps in the side walks

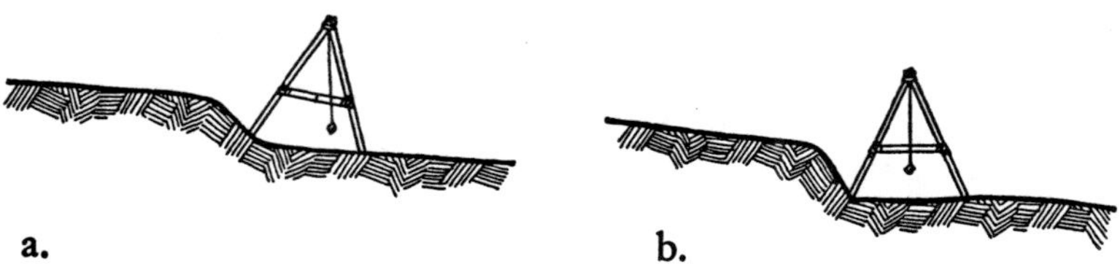

a. b.

Fig. 58 Terrace construction in a slopy area
a. Before b. After construction

Contour-ridge construction

1. Form a wall of stones, logs and/or garden waste on a line marked out with the A-frame. Plant just upslope of the ridge, a row of plants like fruit trees or nitrifying trees like *Leucaena* or any other kind of hedge (Fig. 59).

2. Form a wall of soil and plant, either upslope or on top of it, plants like sugar cane in combination with sweet potato (Fig. 60).

3. Form a dense hedge of a fast growing plant that can be harvested without removing the root system of the plant.

All three types of contour-ridge wall formation can be used to protect the garden bed established behind the contour-ridge walls.

Fig. 59 Contour ridge formation with sugar cane and other plants along the boundary

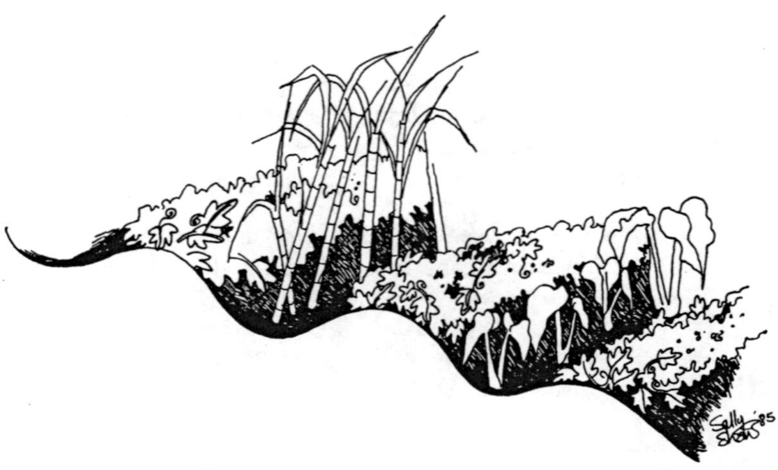

Fig. 60 Contour-ridges in a garden area.

Tree planting system in the garden area

Many different methods exist to plant trees in a garden. Which one to choose depends on the use that you want to make of the trees, and the type of garden in which you want to plant them.

Therefore, only some examples are given together with some remarks about possibilities for the use of the trees.

The following examples will be given:

1. Selected natural regeneration in a traditional garden
2. Improvement of the fallow vegetation of a traditional garden
3. Scattered planting of a mixture of tree species
4. Contour planting in rows and strips
5. East-west orientation of tree planting
6. Orchard tree plantation with underplanting
7. Tree plantation

Selected natural regeneration in a traditional garden

The expression 'natural regeneration' means that you select from the trees in a traditional garden area those which you think are useful.

Trees from which you can make ropes (such as the tulip tree) or trees whose leaves or bark are used for various purposes, such as medicines, should not be eradicated completely. As long as the trees are fairly equally spread over the garden area, it is very unlikely that they will compete substantially for nutrients and space with the planted crops.

As an example, in an area of 25 spade sizes in length and 40 spade sizes in width, about 18 - 36 trees of medium size are adequate.

Improvement of the fallow vegetation of the traditional garden

In addition to the selection of already existing trees in a garden area you can interplant trees that are able to improve the soil, like nitrogen-fixing trees (e.g. *Leucaena* species), or fast growing trees for firewood (e.g. *Leucaena* species).

Scattered planting of a mixture of tree species

This system very closely resembles the system used to improve the traditional garden. In kunai grass gardens, however, very few useful trees get the chance to come up under natural conditions because of the regular burning practises. If you want to establish a garden in these areas make sure that the trees are not damaged or killed by fire. Of great interest are fruit-trees, nitrogen-fixing trees (e.g. *Leucaena* species), or trees for medicinal purposes (e.g. Pawpaw).

The distribution of the trees should be so that the crop you want to plant can easily be planted and harvested.

Scattered planting can also take place in a permanent garden. Before you start to plant the trees you should consider your cultivation system (e.g. rotation system), the futural crops, growth habits of your crop and the growth period of the trees.

If mainly fast growing species are used which will be cut regularly, the number of trees planted can be approximately 20 for an area of 25 spade sizes in length and 40 spade sizes in width.

Contour planting in rows and strips

After having marked the contour with the help of the A-frame we plant trees along the marked line. By this means we will get rows of trees along the contour lines of our garden. The roots will help to prevent soil erosion and their leaves and branches can be used as mulch or give additional compost material.

One way is to plant fast-growing species very narrowly spaced. If they coppiece well they can be regularly harvested.

Another way is bending the young trees and if properly trimmed a good hedge can be made (Fig. 61). Such a hedge of bent down branches of trees is an advantage, particularly under the following conditions:

Fig. 61 Procedure to produce a proper hedge by bending the young trees and trimming them properly

1. Seeds or seedlings are not available in large quantities.

2. A large number of single trees is likely to compete considerably with the food crops for nutrients, space and light.

Another method of separating tree-crops and annual crops in a small area is to plant them in strips according to the contour (Fig. 62). Tree-crops and annual crops are planted alternately on the various terraces. The annual crops can be anything you like to grow. The tree-crops can also be any kind of tree, but preferably trees should be planted that improve the soil and do not attract the same pests and diseases as the annual crops. For example the following tree combinations are recommended:

Coffee + *Casuarina* or coffee + *Leucaena*

The fast growing, nitrogen-fixing trees can be harvested every 3 years and other crops can then be interplanted (Fig. 63). Only the cash crop areas, like coffee or cacao are kept for about 30 years at the same site.

East-West orientation of the trees

In this case the trees you want to plant are lined up from where the sun sets (East-West-orientation). This orientation will give less shadow to the interplanted crop than, for example, the contour planting system.

For fast growing species of trees, which are copiced at least once every year, a spacing of 3–4 spade sizes in length between the tree rows and about 1 spade size in length between the single trees seems to do well in the mid-mountain areas.

Fig. 62 Strip planting of tree-crops and annual crops

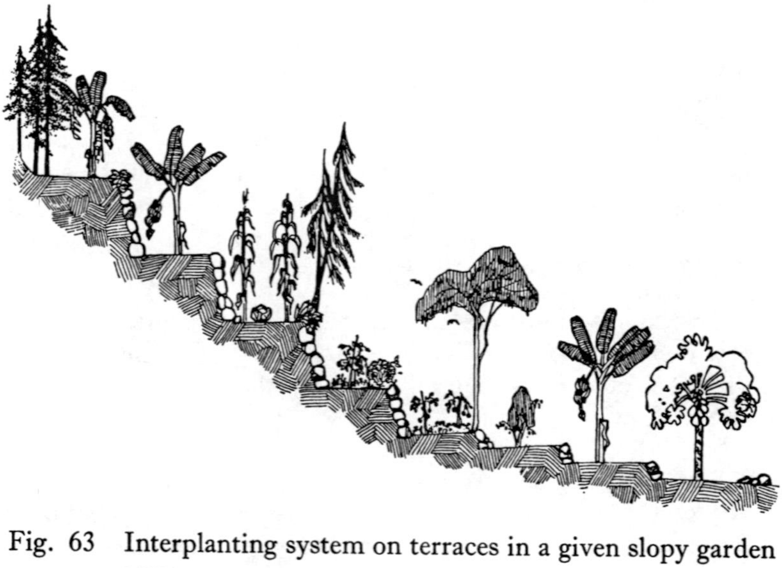

Fig. 63 Interplanting system on terraces in a given slopy garden area

Fruit trees should be planted in rows of at least 7-9 spade sizes distance apart, to be able to grow a substantial yield crop in between the tree rows.

Tree planting or orchard with underplanting

With this method the tree crop is the most important one, be it for fruits, wood or other products. Spacing can be close but not less than 3–4 spades sizes in length between the rows, for fast growing trees and 7–9 spades sizes for fruit trees, unless you grow shade tolerant crops like chinese taro, ginger or cardamon.

Tree plantation

Tree products such as wood, rubber and coffee are the only tree planting products produced in this mid-mountain areas. Interplanting with bananas and papaya is not uncommon. Rubber and coffee need fairly good growing conditions. In areas less suited for agricultural crops, where it is too steep or likely to erode if exposed to heavy rains, tree plantations for firewood and poles should be established.

Whatever system of tree planting is used, regular burning is the greatest enemy. Regular burning kills the trees and replaces them and the other vegetation by weeds if the area is not immediately cultivated. Therefore, the habit of burning all areas is one of the biggest obstacles to soil improvement and the prevention of soil erosion.

Compost production and use

Compost is called "the gold coming from rubbish". How can we produce this kind of "gold" and what is compost in reality?

Compost is something produced out of all kinds of organic rubbish.

Compost is the major nutrition for trees, vegetables and all other growing plants. Plants are able to use the minerals in the compost to grow better and healthier and to produce more fruits, tubers or leaves.

There are very different ways to produce compost, but the one recommended is the following:

All kind of organic waste can be transferred into compost e.g.

— all organic kitchen waste
— all organic garden waste
— leaves from trees
— banana trunks and leaves
— leaves from the bush and all other kind of rotting material
— all kinds of grass
— all kinds of animal manure: this material is of great value, especially to start the composting process

All of these materials form what we call compost. In order to get good, fine compost in a reasonably short time all the collected organic material should be cut with the bushknife into small pieces (see Fig. 64).

Big particles do not rot as quickly and easily as small particles. Therefore, it is essential to cut all the material as much as possible. The better you cut, the better is the compost.

After having cut the material you must mix it thoroughly. This should be done by mixing all material into one big heap, adding shovels or handloads of old compost or animal manure as starter to the heap. After that, the mixture is placed into the compost stockade. In order not to transport the mixture a long way, you should process the entire compost making directly in front of the compost stockade. The coming compost should be placed inside the stockade and lined completely with banana leaves. You should apply water to the compost pile to dampen the material. Cover the top of the compost pile inside the stockade with banana leaves.

The rotting process needs different times in different stockades:

Walkabout stockade	:	2–3 weeks
Banis stockade	:	3–4 weeks
Wire stockade	:	3–4 weeks
Plastic stockade	:	3–4 weeks

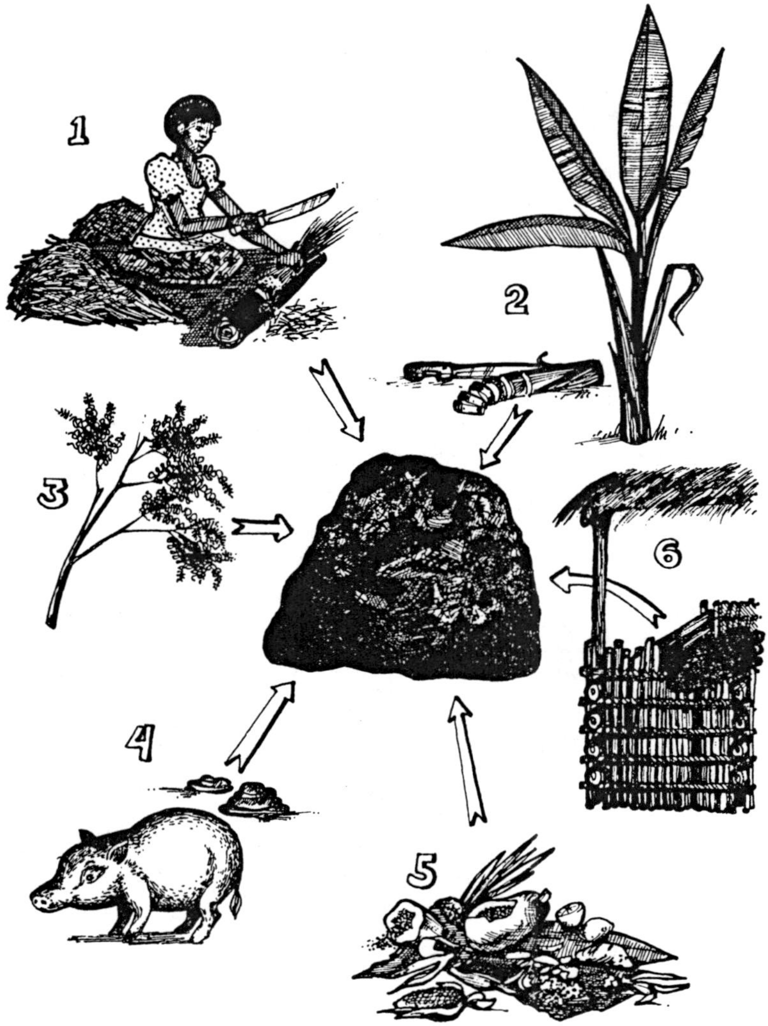

Fig. 64 Compost production
1 Cutting of organic material, 2 Banana trunks, 3 Leaves and the branches of shadow trees (e.g. *Leucaena*), 4 Manure from animals (only necessary as starter, if nothing else is available), 5 Household organic kitchen waste, 6 Old compost as starter

After the decomposition time given above, the compost is black and can be used directly in your garden. The entire composting process should never produce a bad smell if a proper stockade is used.

How to place the compost in your garden?

First you have to prepare the garden beds. Throw a layer of compost on top of the garden bed soil and spread it evenly. After that you have to work the compost with a spade into the soil of your garden bed, so that the soil is afterwards on top of the compost (Fig. 65). The method to transfer the compost under the topsoil is called the "Reverse Dig Method". This method is a great way to till garden beds and it especially benefits deep rooted crops like tomatoes, eggplants, beans and squash.

Fig. 65 Reverse dig method

1. With the spade, dig one trench before you put compost, from one corner of the bed to the other and pile the soil just outside the plot. This soil will be used to fill the final trench.

2. Into the first trench put the compost + topsoil of the second trench by turning your spade accordingly.

3. Dig the next strip and invert the soil + compost into the second trench and so on till the entire bed is ready.

By covering the compost in this way, rain can not remove the valuable plant nutrients remaining in the compost. Garden beds prepared like this are well prepared to yield optimal vegetables.

Planting and protection of seedlings

Different seedlings have to be planted in different ways. The planting method depends on the type of the transplant, its water requirements, the soil and the topography.

The following planting methods are recommended and practised by the Subsistence Agriculture Improvement Project of the Wau Ecology Institute.

Planting of vegetables

Sometimes seedlings of vegetables are transplanted from the nursery to the garden bed without any soil attached to their roots. In these cases it is very important to water them before they are transplanted. During the removal of the 10–15 cm high seedlings from the soil of the nursery the roots should not be damaged. The best time to transplant is the late afternoon on a cloudy day, because then the seedlings will not dry out quickly. At no time should the seedlings be exposed to direct sun or wind.

The plant hole in the garden bed should be big enough to let the roots go down into the ground without bending them (Fig. 66).

Fig. 66 Planting system of vegetable seedlings

Preferably, the seedlings should be transplanted with a clod of soil still attached to the roots. This is easy to obtain with the soil or compost brick system or if the seedlings are placed in banana-fibre-pots.

As with all transplanting, the seedlings should be planted slightly deeper in the garden beds than they were in the nursery. They should be watered immediately after planting and afterwards daily for at least a fortnight. The spacing of most of the common vegetables varies according to the size of the plants during maturation. Common space distances are:

Two stretched hands for all kinds of leaf vegetables like cabbage, aibika, chinese cabbage and one spade size in length for plants like tomato, capsicum, chillies, egg plants, winged beans, yams, taro, cassava and amarathus between the rows and from thumb length to little more than half a spade size of length within the rows. Of great importance is the soil condition in the garden bed. Therefore, the compost application by the "reverse dig method" is very essential.

Planting of banana

Bananas should not be planted on slopes or near the edge of slopes, because they tend to hang over when they bear fruits and it is very likely to uproot the plants if not strongly supported. Preferably you should dig a big enough hole of $\frac{1}{2}$ a spade $\times$ $\frac{1}{2}$ a spade $\times$ $\frac{1}{2}$ a spade, apply some compost and plant the banana sucker in the hole. A wide terrace around the plant will support the growth by collecting and channelling the water run-off to the roots of the plant.

Planting of coconuts and betelnuts

Palms need a lot of water and nutrients. In mid-mountain areas, where rainfall can be irregular, the seedlings should be planted in a big enough hole, so that rain water cannot run off and the silt is transported to the plant. Therefore, a hole of usually $\frac{1}{2}$ a spade depth should be prepared, adding a layer of compost and black soil (Fig. 67). After placing the nut + seedling, the hole is filled to about $\frac{1}{3}$ from the top of the nut. All water run-off and silt will then be trapped in the hole and supply sufficient nutrients to the plant. In very wet areas the holes should not be too deep, because too much water can retard plant growth.

Fig. 67 Planting system of coconuts and betelnuts

Planting of other trees

For most of the trees it is necessary to dig a good sized planting hole. If plants are not taller than about ½ a spade size in length, a hole of approximately ½ a spade in depth, ½ a spade in width and ½ a spade in length is appropriate. After adding two handfuls of compost and some black soil, the seedlings are planted and the soil around them is slightly pressed to support them.

The seedlings transplanted can be:

1. Plants without soil attached to the roots
2. Plants with clods of soil attached to the roots
3. Plants raised in pots, preferably banana-fibre-pots

The tree-plants raised in banana-fibre-pots are easy to transplant and to transport and will usually grow best. In this case the plant + the banana-fibre-pot is planted into the ground.

On slopes, individual terraces should be made around the trees (Fig. 68). Also steps as planting places are good supports for the tree-seedlings (Fig. 69).

Fig. 68 Individual terracing of a papayatree

Protection of vegetable seedlings

Individual seedlings can be protected for the first days from the sun and heavy rains by constructing a small shelter from banana skins (Fig. 70). For this purpose banana trunk skins of about $\frac{1}{2}$ a spade size in length should be cut as shown in Figure 71. The skins should then be folded and fixed to protect the seedlings.

Protection of tree seedlings

For longer seedlings of trees (like coffee), small shelters covered with kunai grass or branches should be used (Fig. 72).

Fig. 69 Supporting system of tree planting by digging steps in a slopy area

Fig. 70 Protection of seedlings by banana-trunk-shelters

Fig. 71 Proper cutting of a banana-trunk-shelter for seedlings

Fig. 72 Shelter system for young trees

Plants for the mid-mountain subsistence garden

The subsistence gardener in the mid-mountain area between 800 and 1800 m above sea level is certainly limited by climatic factors. On the other hand this area is one of the most suitable vegetable growing regions and a vast variety of other valuable tree and foodcrops can be successfully grown.

In the following chapters we will describe the most suitable traditional and introduced vegetables and fruit and nut-producing plants by providing information about their origin, food value, altitudinal limits, growth and planting requirements and possible pests and diseases.

Altitudinal limits of crops

The following table shows the altitudinal limits of the crops which usually can be found in traditional and improved subsistence gardens.

FRUITS

Name	Altitudinal limit in meters above sea level
Avocado	0 – 2200
Lemon	0 – 2100
Mandarines	0 – 2200
Oranges	0 – 1800
Pomelo	0 – 1400
Guava	0 – 1800
Malay Apple	0 – 1200
Mango	0 – 1800
Pandanus (Marita)	0 – 2000
Purple Passionfruit	750 – 2300
Yellow Highland Passionfruit	1300 – 2300
Pawpaw	0 – 1600
Soursack	0 – 1000
Pineapple	0 – 1700

NUTS

Peanuts	0 – 2000
Betel nut	0 – 1000
Coffee arabica	600 – 2100
Coffee robusta	0 – 550
Galip	0 – 700

TRADITIONAL VEGETABLES

Name	Altitudinal limit in meters above sea level
Banana	0 – 2100
Cassava	0 – 1800
Corn	0 – 2500
Sweet potato	0 – 2700
Sago	0 – 1200
Sugar	0 – 2600
Taro	0 – 2700
Chinese taro	0 – 1900
Yam	0 – 1900
Aibika	0 – 1800
Amaranthus	0 – 2400

INTRODUCED VEGETABLES

Mung beans	0 – 1600
Green beans	400 – 2400
Head cabbage	700 – 2400
Capsicum	0 – 2400
Carrots	400 – no limit in PNG
Cauliflower	1000 – no limit in PNG
Celery	900 – no limit in PNG
Lettuce	700 – no limit in PNG
Onion	1000 – no limit in PNG
Peas	900 – no limit in PNG
Pumpkin	0 – 2400
Rubarb	800 – no limit in PNG
Soybeans	0 – 2200
Tomatoes	800 – 2400
Watermelon	0 – 1200
Winged bean	0 – 1800
Cucumber	0 – 1900
English potato	600 – 2700
Chillies	800 – 2000
Ginger	0 – 2000

Traditional subsistence garden in a montane rain forest area. The area has been cleared by the "slash and burn" methode. Some of the bigger trunks remain in the garden site.

Site-stable subsistence garden with contour terraces. Interplanting of tomatoe and marigold prevents nematode attacks.

Subsistence farmer with digging stick

Planted traditional subsistence garden with taro, sweet potato, banana and sugar cane

Triparted compost stockade

Process of mixing chopped greens ready to be placed in the first cell of the triparted compost stockade

Small round "banis"-stockade for compost production in a village

Demonstration of a self-made A-frame, needed to find the points of the same altitude in any given area, before a garden terrace is constructed

Production place for banana-fibre-pots

Banana-fibre-pots planted with young Guava-trees

Soil-steamer to sterilize a soil-compost-mixture needed for the plant nursery

Soil-compost-brick for the breeding of seedlings

Seed drier. The black wrapped part contains rags for the seeds. The clear wrapped part with a layer of stones produces dry hot air.

Self made insert made from bamboo in a given cooking pot, giving the cooking process the advantages of a steam cooking procedure

Various kinds of low-fuel stoves made from different materials

Drum-oven constructed by the Wandumi Women's Association at Wandumi village, Morobe Province, Papua New Guinea

Sweet Potato

Fig. 73 Sweet potato *Ipomoea batata* (L.) Poir.

ORIGIN

Sweet potato probably originated in South America where it has long been cultivated. It is generally thought to have been introduced into Papua New Guinea during the last 400 years.

FOOD VALUE

The main food value of sweet potato is energy, from sugars and starch. It also contains vitamin C and a small amount of protein. The yellow varieties are more valuable nutritionally as they contain large amounts of Vitamin A. Unpeeled sweet potatoes have a higher nutritional value than peeled ones. The young green leaves are a protective food, rich in vitamin A, iron and potassium.

ALTITUDINAL LIMITS

Sweet potato grows well in the lowlands and in the highlands, as high as agriculture is practised. It has been recorded as high as 2850 m.

SOIL TYPE AND CULTIVATION

Sweet potato grows well in most soils, but prefers light sandy soils. It needs a reasonable amount of rain and the yield can be increased substantially if compost, coffee pulp or manure is applied.

Compost should be worked into the soil by the reverse dig method. If coffee pulp is used about one bucket holds enough pulp for about an area of 2 spade sizes in length and two spade sizes in width. Planting on ridges or in moulds is recommended if the soils are heavy. In light sandy soils moulding or ridging is not as important.

SPACING, TIME OF PLANTING AND MATURITY

Planting density is not too important. If ridges are used, two rows are planted per ridge with two cuttings per position. Ridges are usually 1 spade size in length apart and spacing between the cuttings on the ridge is ½ a spade size in length. The cuttings should be about 2 hand spans long (30–40 cm) and be taken from the youngest part of the vine.

Planting time is throughout the year, if there is enough rain. Tubers can be dug after 6–8 months depending on the altitude of the garden.

PESTS AND DISEASES

Sweet Potato weevil: The adults and larvae of this insect bore into stems and tubers of the plant. This happens more often during dry weather when the soil cracks, allowing the beetle easy access to the tubers.

The best control is crop rotation and good cultivation.

Sweet Potato Hawkmoth: The caterpillars of this insect attack the leaves of the plant.

The best controls are chickens and toads in the garden.

Sweet Potato Leaf Miner: The small larvae of this moth mine into the leaves.

No control measures are recommended.

Little Leaf: Outbreaks are only found in some restricted areas.

Leaf Scab: This is a widespread fungus infection with very little effect on the yield.

No control measures are necessary.

Taro

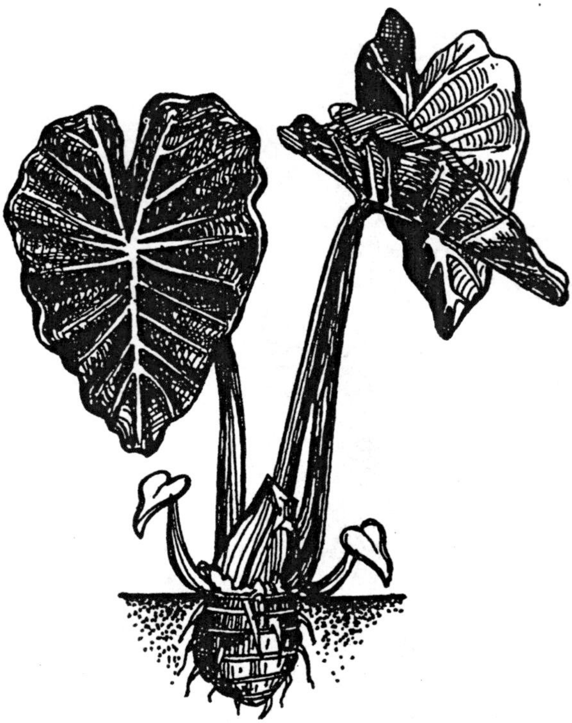

Fig. 74 Chinese Taro *Colocasia esculenta* (L.) Schott

ORIGIN

The origin of taro is thought to be South-East Asia and Central Asia. It was dispersed throughout the Pacific with the migration of Polynesian people from the Malayan Peninsula via the Sunda Islands and New Guinea. It has been cultivated as a food crop for more than 2000 years.

FOOD VALUE

It is mainly grown as energy food according to the starch contents of the corms but it also contains substantial amounts of minerals like potassium, phosphorus and calcium. The leaves are rich in vitamins A and C and contain a high amount of protein.

Fig. 75 Taro kongkong *Colocasia spec.*

ALTITUDINAL LIMITS

Taro will grow best above 1000 m up to 2500 m. Unlike most other crops it is cultivated under both wetland and dryland conditions. Its tolerance of waterlogged conditions is due to the ability of the plant to transport oxygen from the leaves to the roots.

SOIL TYPE AND CULTIVATION

It can be grown in a large variety of soils but will do best in dark coloured soil, rich in organic matter, fairly soft and friable.

The planting depth is between half to one handspan depending on the size of the sett. Usually the planting holes are made using sticks or spades.

Taro is propagated vegetatively using setts from mother plants or suckers. Large and medium sized setts produce higher yields than smaller setts.

SPACING, TIME OF PLANTING AND MATURITY

The average plant spacing is about one spade size between the individual plants. Any time of the year is suitable to plant taro. Usually every time there is a harvest there is a new crop to be planted. The beginning of the rainy season is the most suitable time for planting.

Crop duration for taro varies depending upon variety and soil fertility, but is generally 6–9 months.

PESTS AND DISEASES

Taro beetle: These are the most serious pests of taro in Papua New Guinea. The caterpillars of these black beetles feed on the corms and roots of taro. Rotting logs are the hiding places for the beetles. Therefore these kind of logs should be removed from the sites where taro will be planted.

Taro leaf hopper: These small white or black leaf hoppers cause direct damage to taro by feeding on the petioles and upper parts of the stem. These insects can be controlled by the mirid egg predator that preys on the eggs of the leaf hoppers.

Taro hawk moth: The larvae of this moth are up to 6 cm long and sometimes cause defoliation of the taro. The giant toad is a good biological agent for the control of this pest.

Army worms: Worm larvae feed on the top surface of the taro leaves. They usually feed together in large numbers. A pentatomid bug is a good biological agent to control this pest.

Taro leaf blight: Rotting of the petioles occurs after the foliage is destroyed by this fungus.

Rotation and removal of leaves infested with small, dark, roundish spots is the best counter measure. Intercropping with other crops such as aibika and banana is also a good preventive measure.

Viruses: These viruses cause malformation of plants. Immediate removal and burning of these plants is the best counter measure.

Soft rot: Wilting and change of colour of the leaves is a result of this soil-borne fungal disease.

Rotation and proper handling of the soil is a good counter measure.

Cassava

ORIGIN

Cassava is not known in the wild state, but has long been cultivated in Central and South America. It was probably introduced to Papua New Guinea some time before 1800 and much later came to the highlands via the coastal areas.

FOOD VALUE

The main food value is energy from starch. They contain modest amounts of the minerals calcium, phosphorus, iron and vitamin C. The leaves are high in vitamin A, and relatively high in protein, fibre, calcium and iron.

Small amounts of cyanogenetic chemicals are present and poisonous if the tuber is eaten raw. Normal cooking removes these poisonous substances.

ALTITUDINAL LIMITS

Cassava grows well in a moist, warm climate. It is commonly grown up to 1700 m in the highlands.

Fig. 76 Cassava *Manihot esculenta* Crantz

SOIL TYPE AND CULTIVATION

Cassava grows best in light sandy soil with good drainage. It will give good yields on heavily cropped soils which are unsuitable for other crops, and so is often the last crop in a rotation.

In areas with low, uneven rainfall, cassava is a suitable crop to grow. In times of drought the plant stops growth and drops its leaves. It recovers quickly after rain.

Cassava is usually propagated by thick stem cuttings planted in the soil. Any cultivation method of the soil improves the yield substantially.

SPACING, TIME OF PLANTING AND MATURITY

Sticks with about two buds (¼ a spade size in length) of about 10 months old are best. Planting distance should be about 1 spade size in length.

Any time of the year is suitable for planting, if enough rain is available. Crops may be harvested 10 – 12 months after planting. Tubers become woody if left to long in the ground.

PEST AND DISEASES

The crop appears free of major pests and disease problems here in Papua New Guinea. However, pig and rat damage to tubers can be significant.

Yams

ORIGIN

Most of the more than 60 species known originate from tropical and subtropical Asia, some from Africa.

The yam species *Discorea nummularia* is now principally a yam of the South Pacific including Papua New Guinea.

FOOD VALUE

The main value lies in the amount of starch stored in the tuber, making this plant a good source of energy. A moderate amount of protein is also found.

ALTITUDINAL LIMITS

They will grow almost everywhere in Papua New Guinea up to 1900 m.

Fig. 77 Yams *Dioscorea alata* L.

SOIL TYPE AND CULTIVATION

Loose, deep, fertile soil is best. Yams will not grow well in sandy soils and the drainage and aeration of the soil is important. Compost application will substantially improve the yield. The planting in individual hills or high ridges supports proper drainage and aeration, but stimulates soil erosion.

They are usually propagated from aerial tubers or pieces of tubers. The climbing vines need support e.g. by a strong stake of about 2 spade sizes in height.

SPACING, TIME OF PLANTING AND MATURITY

Spacing should not exceed one spade size between plants in any direction. The time of planting is usually adjusted to coincide with the beginning of the rainy season. But if enough water is available, any time of the year is suitable for planting.

After about 9–10 months the tubers are ready for harvesting.

PESTS AND DISEASES

No serious pests and diseases have been reported for Papua New Guinea.

English Potato

Fig. 78 English Potato *Solanum tuberosum* L.

ORIGIN

The potato originated in Peru and from there its cultivation has spread throughout the world. It was introduced to Papua New Guinea by Europeans.

FOOD VALUE

The potato is lower in food energy than the sweet potato but higher in protein. It is a good source of vitamin C.

ALTITUDINAL LIMITS

In all highland areas of Papua New Guinea potatoes can be grown successfully. The ideal range appears to be 1500 to 2700 m.

SOIL TYPE AND CULTIVATION

A wide range of soil types from clay to sand are suitable, but deep loose soils with decomposted organic matter are best.

It is important that the site has not been planted with other solanaceous crops (e.g. tomato, egg plant, capsicum) within the previous two years, to guard against outbreaks of bacterial wilt.

Well composted moulds is the best cultivation method. The soil should be broken at least 10 weeks before planting. This will allow time for any organic matter to decompose. Undecomposed organic matter ties up soil nitrogen and causes the skin of the potato tuber to become rough.

The area should be well drained, as potatoes will not grow in waterlogged conditions.

SPACING, TIME OF PLANTING AND MATURITY

Ridges should be formed and two rows planted on each ridge. These rows should be $\frac{1}{2}$ a spade size in length apart. Plants within the row should be $\frac{1}{4}$ of a spade size apart. Planting depth is one spade blade. Seed potatoes should be planted after rainfall the year round. Potatoes will mature in 100 – 120 days. They should not be dug before all leaves and stalks have properly dried out. Before that the tubers are immature. After harvesting, the tubers should be handled as little as possible.

Potato Tuber Moth: The adult moth will lay the eggs on the surface of the leaves or on exposed tubers. When the larvae emerge from these eggs they burrow through the leaves, causing brown blotches. In tubers they burrow narrow tunnels just under the skin. If tubers are not exposed, no damage can be done. Only in the case of very severe leaf mining preventive measures are recommended.

Taro beetle: These are shiny black beetles with horns on the head and thorax. They eat holes in the tubers.

Control measures are usually not necessary.

Other insect pests: This includes cutworms, wireworms and various species of beetles and caterpillars.

Control measures are usually considered unnecessary.

Bacterial wilt: This is the most serious disease of potatoes in Papua New Guinea. In the early stages a few leaves of the infected plants wilt. Later the whole stalk and then the complete plant will wilt and collapse. Usually all of the tubers of infected plants decay.

Control measures are:

- Remove infected plants immediately and burn them.
- Use a long rotation of at least two years to allow the soil to rest after a crop of potato.
- At least two years should pass before potatoes or other solanaceous crops (egg plants, tomato, capsicum) are planted again on the same area.
- Plant only in new clean ground and only with clean seed and clean tools.

Target spot: This is caused by a fungus which produces dark brown spots on the leaves. These spots spread outwards until the whole leaf is covered.

In serious cases spraying with a fungicide is recommended.

Root knot nematodes: The larvae of this slender worm enter the potato tubers.

The best control is crop rotation and the simultaneous planting of marigold plants around the potato fields. Marigold releases a substance into the ground which is disliked by nematodes.

Green Beans

Fig. 79 Green Beans *Phaseolus vulgaris* L.

ORIGIN

Green beans originated in Central America. Spanish and Portuguese explorers took the beans to other parts of the world.

FOOD VALUE

All contain useful amounts of a wide range of minerals, vitamins C, B, and A. Beans show a variation in food value from one variety to another.

ALTITUDINAL LIMITS

Beans can be grown to the limits of agriculture in Papua New Guinea to about 2400 m.

SOIL TYPE AND CULTIVATION

Well drained, compost-rich soil is best. Beans can be planted on the flat or on top of a ridge. For climbing varieties strong stakes must be arranged. Early control of weeds by hand weeding is recommended.

SPACING, TIME OF PLANTING AND MATURITY

Ridges should be formed ½ a spade size apart. Planting should be done along these ridges with one handspan between seeds.

Any time of the year is satisfactory for planting, provided enough water is available.

As beans are grown for their vitamin content they should be harvested while still thin. Plants can be kept in production for 1–2 months by picking the right sized beans every seven days.

The pods should start reaching maturity after 8–10 weeks.

PESTS AND DISEASES

As the area of beans being sown is likely to be small, it is better to accept a small amount of insect damage and avoid any treatment with chemicals. Bean fly and Bean Pod Borer are main insect pests in Papua New Guinea.

Winged Beans

ORIGIN

It probably originated in tropical Asia, perhaps Java. It has long been known to Melanesians including Papua New Guineans.

FOOD VALUE

The main food value of winged bean is protein. Almost all parts can be eaten: tubers, the young leaves, the flowers, the young pods and the seeds. Young pods are the most popular part of the crop. Fully riped seeds are only used for planting.

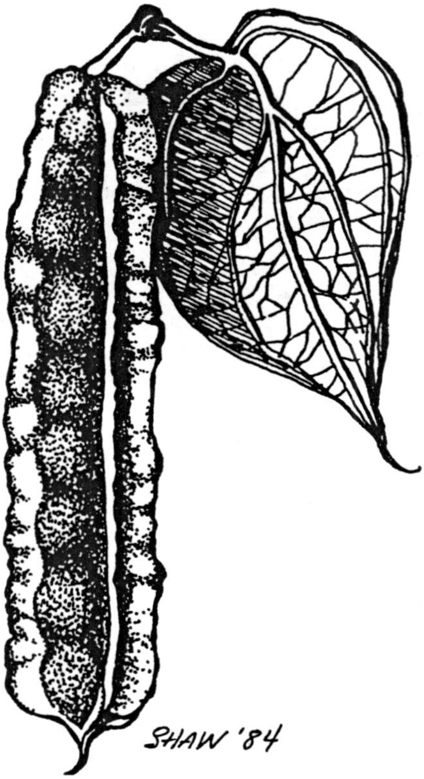

Fig. 80 Winged beans *Psophocarpus tetragonolobus* DC.

ALTITUDINAL LIMITS

They will grow almost everywhere in Papua New Guinea up to 1800 m.

SOIL TYPE AND CULTIVATION

Any compost enriched soil is good to grow winged beans. Well rotted coffee pulp substantially increases the tuber yield if added at $\frac{1}{2}$ a bucket per two spade sizes in length and width (2 square meters).

Hand weeding during the first 6 weeks encourages the healthy growth of the plants.

Winged beans are able to enrich the soil with nitrogen. They are capable of improving the quality of the soil, if the roots are left to rot in the soil after the plants have been harvested. By planting winged beans as part of rotation, nitrogen, which has been used up by preceding crops, is returned to the soil.

SPACING, TIME OF PLANTING AND MATURITY

On ridges, the holes for the seeds which are made with the finger, should be about a handspan apart. The single ridges should be about ½ a spade size apart. The plants need to be helped to climb off the ground. Therefore stakes (about one for two to three plants) should be put into the ground so that the plants are able to climb. The stalks should be about 1½ spade sizes in length.

The beginning of the dry season is the best time to plant. Fewer flower pests and diseases attack the plants.

The maturity of the pods varies considerably from place to place. In the lowlands about 10 weeks and in the highlands about 12 weeks are needed for pods to be ready for harvesting.

Seeds will be ready for planting or eating after about 4–5 months.

PESTS AND DISEASES

False rust: It causes small coloured lumps on the leaves, stems and pods. The whole pod may turn purple.

Powdery mildew: It looks like white powder on the upper surface of the leaf. Both diseases can be controlled with fungicides but it is too expensive to be recommended. Planting during relatively dry periods and planting the winged beans in a different place each time, using new stakes every time is a much better method of control of these diseases.

Root knot nematodes: These tiny worms live in the soil and cause lumps on the tubers. Planting marigold flowers in the vicinity of the winged beans will help to prevent destruction by nematodes.

Cabbage

Fig. 81 Cabbage *Brassica oleracea* L. var. *capita* L.

ORIGIN

Cabbage has been known in Europe for more than 4000 years. It came with the Europeans to Papua New Guinea.

FOOD VALUE

The main food value in cabbage is vitamin C. This vitamin is destroyed by cooking. For the best value, cabbage should be eaten slightly crisp.

ALTITUDINAL LIMIT

Cabbage will grow in the mid-mountain areas and in the highlands up to 2400 m.

SOIL TYPE AND CULTIVATION

Rich sandy soil is best. The addition of compost will considerably improve the harvest.

Seeds are sown first in large banana-fibre-pots and later transplanted. When planting, a deep hole must be dug so that the tap root of the young plant remains straight. The seedlings must be well watered after planting.

SPACING, TIME OF PLANTING AND MATURITY

The seeds, if planted in a nursery should be sown in so that after germination there is a distance of about one finger length between the young plants. Raising in banana-fibre-pots is much easier. The number of seeds per pot depends on the size of the pot. The young plants can be transplanted after about 6 weeks.

For planting, the distance between rows and between plants in a row depends on the type of cabbage being grown. Small varieties may be grown one handspan apart with one handspan between the rows. Large cabbage heads need two handspans between plants and about three handspans between the rows.

The planting of the seedlings can be done at any time of the year. The best planting time is the late afternoon on a cloudy day to avoid a too rapid wilting of the seedlings during re-planting. Depending on variety and climate, cabbage will take 2–3 months to mature. Harvesting should be done when the cabbage heart is firm.

PESTS AND DISEASES

Diamond back moth: The very small caterpillars of this moth which are pale brown at first and become green when fully grown, eat holes in the leaves.

Cabbage cluster caterpillar: These are small yellow-green caterpillars with white stripes and black dots on the body. They feed in the center of the cabbage.

Soil insects: They include cut worms, crickets and the taro beetle.

As a control measure, Derris dust is recommended instead of the variety of poisonous and expensive chemicals. Derris dust can be used up to 3 days before the harvest. Dusting should be done in the early morning, when the leaves are wet and when there is no wind.

Do not breathe in the Derris dust while spreading it.

Aibika

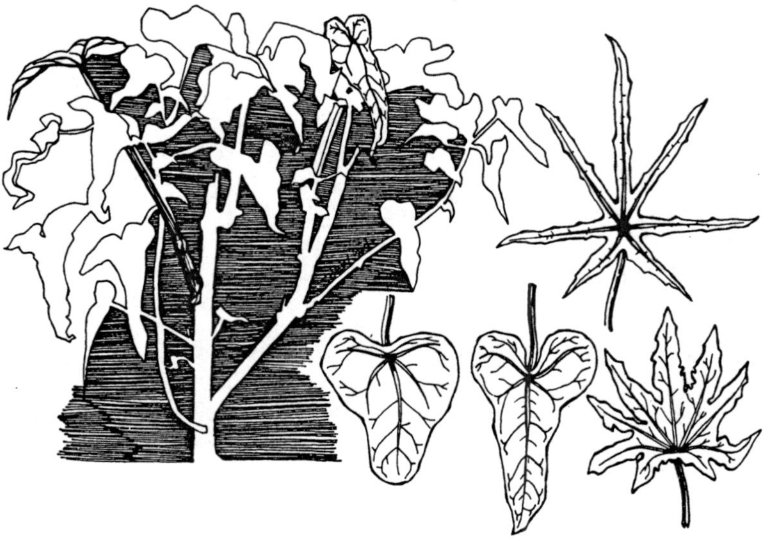

Fig. 82 Aibika
The variability of Aibika leaves is shown

ORIGIN

Today it is widely distributed throughout the islands of the South Pacific. It is only little known in the Western Hemisphere.

FOOD VALUE

The most important factor is the high content of potassium in the leaves together with substantial amounts of vitamin A.

ALTITUDINAL LIMITS

It can be grown in most areas which are not waterlogged.

It is hardly growing in grassland areas.

SOIL TYPE AND CULTIVATION

Aibika or 'sunset hibiscus' does not require any specific soil type but does best in compost-rich soil.

The crop is a well known shrub of about 1 spade size height. It is propagated from stem cuttings. The leaf shapes are variable due to a great variety of Aibika species available. It usually grows for years with a minimum of care.

SPACING, TIME OF PLANTING AND MATURITY

No special requirements for spacing are necessary and the plants can be planted at any time of the year. Leaves and young pods can be collected about 2–3 months after planting and the production continues for 1–2 years.

PESTS AND DISEASES

Beetles and caterpillars of moths can cause severe damage to the foliage. The Stem borer also can cause serious damage to the plants.

Intercropping and rotation is a good counter measure to reduce this sort of damage.

Water Leaf

ORIGIN

It is a common plant in all tropical areas and is sometimes regarded as a weed.

FOOD VALUE

It contains mainly vitamin C and calcium is found in large quantities in the leaves of this plant. The taste is slightly sour due to oxalate (mineral) which is found in the green parts of the plant.

ALTITUDINAL LIMITS

It will grow up to 2000 m but does best between 800 and 1500 m.

Fig. 83 Water leaf or Kumu manus *Talinum triangulare* (Jacq.) Willd.

SOIL TYPE AND CULTIVATION

Talinum can be grown in rather infertile soils and during all seasons. It is propagated by seeds and occasionally by cuttings. It has a remarkable drought resistance. Seed collection is not easy because the seeds are very small.

SPACING, TIME OF PLANTING AND MATURITY

No special spacing arrangements are required. It can be planted at any time of the year. The cutting of edible shoots may continue for more than a year.

PESTS AND DISEASES

No serious pests or diseases are known.

Amaranthus

Fig. 84 Amaranthus *Amaranthus tricolor* L.

ORIGIN

Seven centers of origin are recorded for this plant. They are located in East Asia, South-East Asia, India, the Mediterranean, South America, Central America and North America.

FOOD VALUE

The leaves are a fair to very good source of protein and the carbohydrate content is also fairly high. Vitamins A and C are found in high concentrations. Of the minerals, calcium and iron are plentiful.

ALTITUDINAL LIMITS

Amaranthus is a warm-season vegetable adapted to the conditions of the hot, humid tropics but is also suitable for temperate zones during the summer. They can be grown everywhere in Papua New Guinea up to 2400 m.

SOIL TYPE AND CULTIVATION

These plants will grow in a great variety of soils but it does best in soils to which large amounts of compost have been added. For best growth, soil fertility should be high.

The small seeds of *Amaranthus* must be planted shallowly, covered with fine soil or sand and watered frequently in order to establish healthy plants. The best way to germinate the seeds is in a highly fertile seedbed. Four to five weeks after seeding the plants can be transplanted to the field.

SPACING, TIME OF PLANTING AND MATURITY

A distance of about half a handspan between the single plants is appropriate. Transplanting should be done on a cool day or during the cool hours of the late afternoon, and the plants should be carefully watered.

They will be ready to harvest about 3 weeks after transplanting. Control of water is very important. The garden beds should not be flooded.

The leaves can be harvested any time after planting. The seedlings are a very good vegetable. The most common practise is to harvest *Amaranthus* by repeated cutting, which substantially increases the yield.

PESTS AND DISEASES

Pests that attack *Amaranthus* vary from region to region, but diseases are few. Only some caterpillars may cause damage to the plants.

Pumpkin

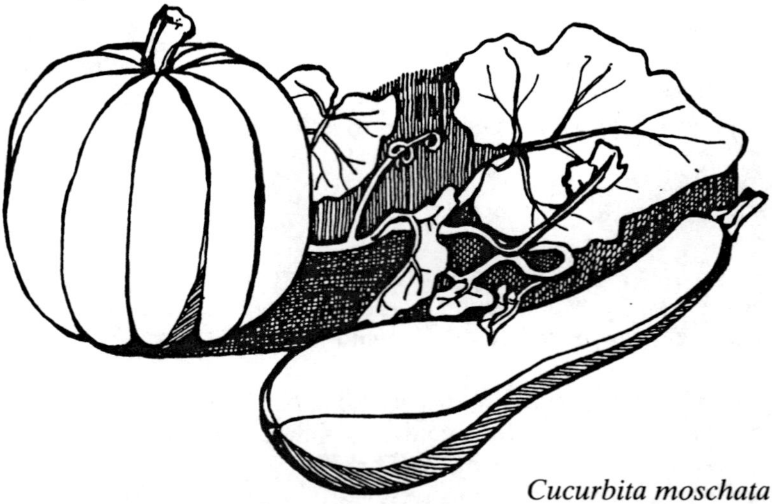

Cucurbita moschata (Duch.) Poir

Fig. 85 Pumpkins of various shape

ORIGIN

It appears that pumpkin was domesticated in South America. It is now grown in most tropical, subtropical and temperate areas of the world.

FOOD VALUE

Pumpkin tips, leaves and fruit are particularly rich in vitamin A. All parts of the plant have moderate amounts of vitamin C.

ALTITUDINAL LIMITS

Pumpkin will grow throughout the highlands to an altitude of about 2400 meters.

SOIL TYPE AND CULTIVATION

Pumpkin grows on a wide variety of soil types but does particularly well in soils with plenty of decaying organic matter.

Vines can be grown on ridges or on moulds. Mulching is commonly done after the application of compost and manure.

SPACING, TIME OF PLANTING AND MATURITY

Planting on ridges should be done two spade sizes apart with one spade size spacing between plants in a row. Each plant requires approximately 2 spade sizes in a square but if left, vines will spread throughout the garden and cover other vegetables.

Pumpkin can be planted at any time of the year. The pumpkin fruit will start maturing about 3–4 months after planting, although the green leaves and tips will be ready for eating well before this time. One pumpkin vine will continue to produce for up to a year.

PESTS AND DISEASES

There are no pests or diseases that cause serious crop losses, although mildew can sometimes be a problem.

Onions

ORIGIN

The probable origin is from the subtropical areas of the world.

FOOD VALUE

Its main food value is in the oily sap which is a very tasty addition to soups and any kind of vegetable diet.

ALTITUDINAL LIMITS

Onions do not grow well in the lowlands and they will not produce bulbs under tropical conditions, because the minimum day length must be 13 hours. Nevertheless they do best from 1500 m to 2400 m.

SOIL TYPE AND CULTIVATION

A compost rich soil is most suitable for onions and adequate moisture at all times is important.

Fig. 86 Onions *Allium sativum* L.

SPACING, TIME OF PLANTING AND MATURITY

Onions can be sown directly into the field or transplanted from a nursery. When the plants are about one handspan tall they are ready to be transplanted. The distance between the rows should be about one handspan and the space between the individual plants should be about half a handspan.

They can be planted the whole year round.

The plants need 18–30 weeks to mature, depending on the growing conditions, the altitude and the variety transplanted.

PESTS AND DISEASES

Mildew is one of the most serious diseases. The use of resistant varieties such as Hybrid Granex White is the best solution.

Tomato

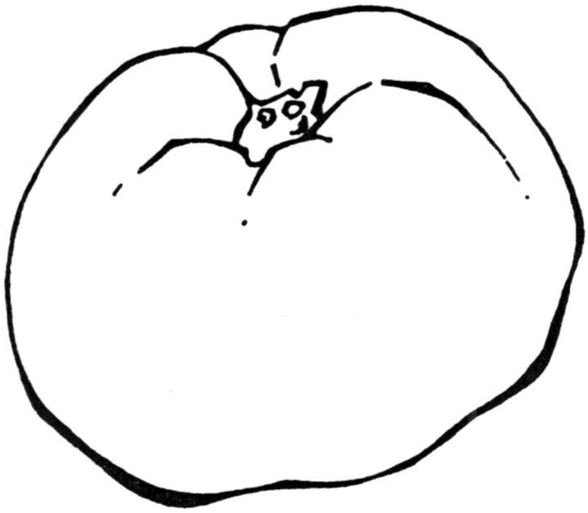

Fig. 87 Tomato *Lycopersicon esculentum* Mill

ORIGIN

It originates from tropical America and has spread throughout the world.

FOOD VALUE

A very high content of vitamins A and C is the main food value of tomato fruit.

ALTITUDINAL LIMITS

The temperature should never be lower than about 18°C. Otherwise the setting of fruits will be affected. Therefore, tomatoes grow best between 800 and 1500 m.

SOIL TYPE AND CULTIVATION

A compost rich soil best supports the growth of these plants.

Cultivation can be done throughout the year, however the rainy season is usually considered difficult for growing tomatoes due to diseases.

SPACING, TIME OF PLANTING AND MATURITY

The seedlings should be raised in banana-fibre-pots. When the seedlings are half a handspan tall they are ready to be transplanted. The time of planting can be any day throughout the year, but it should be a rainy or cloudy day and the time to transplant should be late afternoon. Careful watering and protection of the seedlings from sun and torrential rain is very important (see Fig. 70). By covering most of the stem with soil, more roots will grow. This enables the plant to grow better and more quickly.

The distance between the rows should be one spade size and the space between the individual plants in the rows should be half a spade size.

In about 14 – 18 weeks the plants should have harvestable fruits.

Tomatoes are not strong plants, so they need to be tied up with string to a strong stick placed in the ground beside the plant. This also makes it easier to pick the fruits. Side shoots should be cut off leaving only one or two main stems. This will result in fruits of a much better quality and quantity.

PESTS AND DISEASES

During the rainy season, blights, fruit rot, moulds and leaf spots are especially prevalent.

Tomato stem borer: This caterpillar attacks the inside of the stem which causes the plant to wilt. Rotation of the crop plants is a good preventive measure as well as intercropping, e.g. with peanuts, kumu manus or cabbage.

Nematodes: These worms attack the roots and cause the plant to wilt. The planting of marigold in the vicinity of the tomatoes will effectively prevent nematode attacks on the tomatoes.

Capsicum

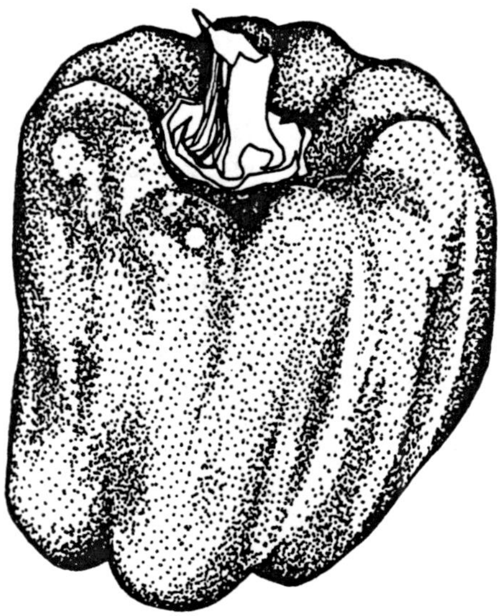

Fig. 88 Capsicum or Green Pepper *Capsicum spec.*

ORIGIN

It originates from tropical and subtropical America and has been known for more than a thousand years. They were adopted widely. The slightly pungent varieties are preferred in the USA, Europe and Papua New Guinea, while the very pungent peppers (e.g. chillies) are indispensable in other countries, such as India and parts of Africa.

FOOD VALUE

The main food value in capsicum is vitamin A. There are also large quantities of phosphorus and calcium found in capsicum.

ALTITUDINAL LIMITS

Peppers are warm-season crops that can be grown at any time of the year in the hot, humid tropics.

SOIL TYPE AND CULTIVATION

Peppers can be grown in any soil but they prefer a fertile, loamy soil.

Seedlings are grown first in larger banana-fibre-pots and then transplanted to other flats of soil once they have produced two true leaves.

Plants must not dry out during transplanting and the young plants need to be watered regularly.

When the plants are half a handspan high they can be exposed to progressively drier and sunnier conditions and to high and low temperatures.

SPACING AND TIME OF PLANTING AND MATURITY

Spacing in the field may vary considerably, but about one spade size between rows and about one handspan between plants within a row are convenient distances.

Seedlings transplanted directly in the field are often subject to sever shock when planted. Therefore planting should be done on as cool a day as possible or in the late afternoon hours. Immediately after transplanting the plants should be watered carefully.

Flowering begins six weeks after seeding and continues for about 4 months. Old plants that have lost most of their leaves are no longer productive and should be used for composting.

PESTS AND DISEASES

Viruses: Symptoms of viruses including mottling, twisting and distortion of leaves, reduced leaf width or size, and stunted growth.

Rotation and avoiding other possible hosts such as tomatoes or tobacco on the same piece of land is a biological control measure.

White mite: Symptoms are bronzing of the leaves or twisting of the leaves.

To avoid this, seedlings should be grown in non-contaminated areas.

Bacterial wilt: In these cases the lowermost leaves turn yellow and wilt.

Good drainage helps to prevent the disease and contaminated plants should be removed.

Insect pests: The pepper weevil can cause severe damage.

Rotation practises are the best control measure to prevent serious damage by insects.

Sugar Cane

ORIGIN

This important plant originates from New Guinea and has been distributed eastwards and westwards throughout the tropical and subtropical world.

FOOD VALUE

The plant is known for the high amount of sugar and is therefore a very good source of energy.

ALTITUDINAL LIMITS

Sugar cane can be grown up to 2600 m but does best up to 1000 m above sea level.

SOIL TYPE AND CULTIVATION

Any type of soil rich in compost or any other organic matter will suit the growth of this plant.

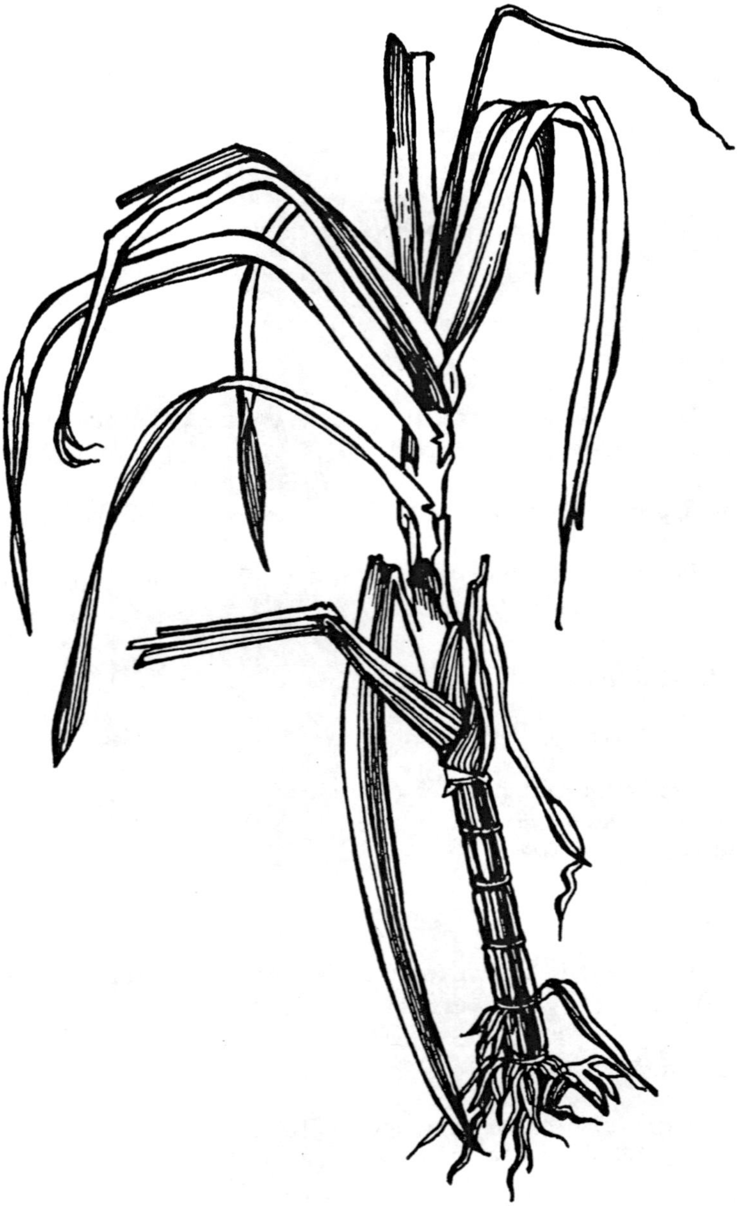

Fig. 89 Sugar Cane *Saccharum officinarum* L.

The planting material is stem cuttings (Fig. 90) or setts with 2–4 nodes each.

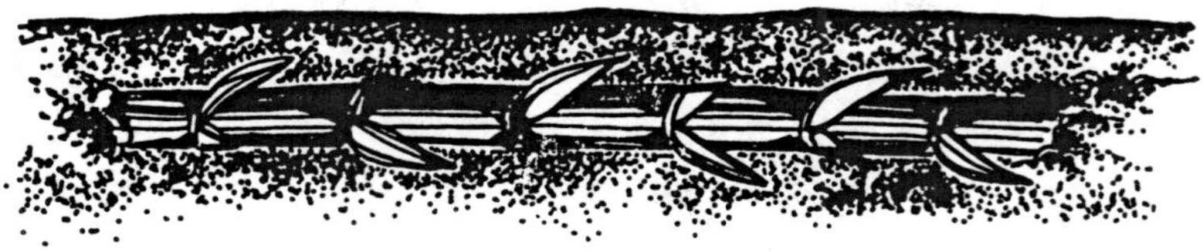

Fig. 90 Correct planting of sugar cane. The "eyes" should lay horizontal.

To build up a substantial amount of sugar in the cane, a natural period of stress is required; usually a dry period.

SPACING, TIME OF PLANTING AND MATURITY

The setts should be planted about half a spade length apart and in rows with one spade length distance in between. The time of planting should be chosen so that the young plants get enough rain and so that before the first cutting, about 12–18 months after planting, there will be a dry period. The plants are by then 2–3 spade sizes in height.

When the stems appear woody they are ready to be cut. The successive stems, called "ratoons", need approximately the same time to mature as the first plantings. After two to three ratoon crops, replanting is usually necessary.

PESTS AND DISEASES

There are no special pests and diseases for sugar cane found in Papua New Guinea, but insects and leaf spot fungus can attack the plants and cause serious damage.

Pit Pit

ORIGIN

The origin of this cane is not yet fully known. There are some indications that it might be a mutant derived from sugar cane. The original wild types have not yet been found. If the

Fig. 91 Pit Pit *Saccharum spec.*

garden types are neglected and not propagated vegetatively, the types usually used in subsistence gardens seem to die out because they usually do not produce seed. This supports the theory that pit pit might be a mutant of sugar cane.

FOOD VALUE

Pit Pit has no important food value but contains a moderate amount of protein in the edible portion.

The tender inflorescence makes it a tasty vegetable whether cooked or eaten raw.

ALTITUDINAL LIMITS

The plant can be grown throughout Papua New Guinea. The area around Mt. Hagen has a special variety of pit pit known as "Moi". Wild forms of this kind of pit pit are known. It is an important famine food as the plant survives in fallow gardens for many years.

SOIL TYPE AND CULTIVATION

Any type of soil is suitable to grow pit pit as long as the soil is not waterlogged.

Pit Pit is grown from cuttings of a handspan in length. Each cutting should contain about 3–4 nodes. About 3–4 cuttings are usually planted in one hole burying about half a handspan of the cuttings in the ground.

Used as boundary plant the pit pit can help in erosion control.

SPACING, TIME OF PLANTING AND MATURITY

There is no specific spacing necessary and the planting can be performed at any time and date of the year.

After 7–9 months the unopened flower on top of the plant can be harvested. Generally it is considered to be ripe when most of the leaves on the stalk are brownish. After about 3–4 years the edible portion becomes smaller and smaller. New planting is then necessary.

PESTS AND DISEASES

Fungus can attack the edible parts of the plant if not harvested in time. Other serious diseases or pests are not recorded.

Corn

ORIGIN

Corn is thought to have originated in Central America perhaps 7000 years ago. It was introduced to Papua New Guinea by Europeans.

FOOD VALUE

Corn is usually eaten when it is almost ripe but while the seeds are still soft. It provides energy in form of starch, some protein and some minerals.

Fig. 92 Corn *Zea mays* L.

ALTITUDINAL LIMITS

It grows well to an altitude of about 2500 m.

SOIL TYPE AND CULTIVATION

Corn can be grown on many soil types, providing the rainfall is adequate and the land is well drained. Addition of compost and proper weeding until the grain is well formed will increase the yield substantially.

SPACING, TIME OF PLANTING AND MATURITY

Seeds should be planted about one finger length in depth and two handspans between the plants and about three handspans between the rows. In areas with high rainfall the spacing is much closer: one handspan between the plants, three handspans between the rows.

Planting time is the wetter season and the harvest is in the drier season. Corn will be fully ripe in about 90-100 days. In the highlands this may take up to 150 days.

PESTS AND DISEASES

Birds: Parrots are one of the worst pests. Damage can be reduced if tins or anything that makes noise in the wind are hung above the corn.

Cutworm: The larvae of this insect live in the soil and chew through the stems of the young plants.

The best control is by rotation of crops.

Army worm: These larvae eat the entire above-ground portion of the plant, or chew through them at ground level.

Crop rotation will help prevent this type of pest damage.

Corn borer: These small larvae bore into the stems of the plants.

There is no measurable yield reduction, therefore no control measures are necessary.

Corn Earworm: The caterpillars feed on the developing kernels at the tassel end of the cob. The damaged tip can be removed. Yield is usually not reduced. Therefore no control measures are recommended.

Rust: This is one of the most serious corn diseases in Papua New Guinea. It causes small brown spots on the leaves and stem.

Two kinds of fungi cause the rust. The best control for rust is to use resistant varieties provided by the Department for Primary Industries (DPI).

Leaf Blight: Both of the two types known in Papua New Guinea produce greyish-brown spots on the leaves.

Blister Smut: Usually only a few plants in the crop are infected. Large blisters filled with a black mass on the cob or stem are formed.

The infected plants should be removed and burned. The area should not be used for corn for about one year.

Downy Mildew: It causes plants to stay small. In severe cases, no cobs will be formed.

The best control measure for all diseases is crop rotation. Avoid planting one crop of corn followed by another crop of corn.

Peanuts

ORIGIN

The peanut originated in South America. It is now found in all tropical and sub-tropical countries.

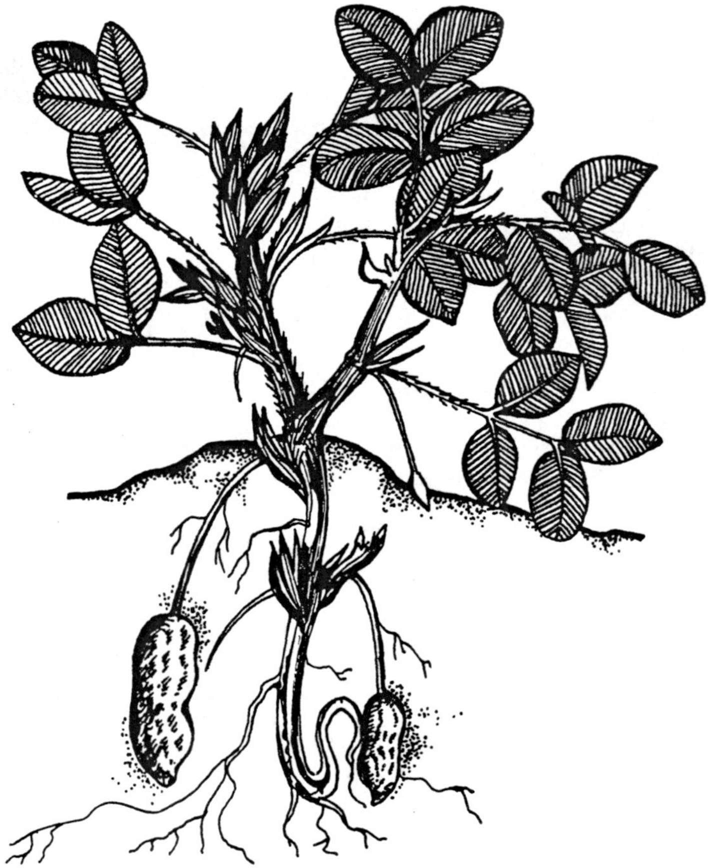

Fig. 93 Peanut *Ararchis hypogäa* L.

FOOD VALUE

The peanuts main food value is protein. As well as protein, peanuts contain alot of fat and therefore provide a good source of energy.

They are also rich in vitamins B and E.

ALTITUDINAL LIMITS

Peanuts grow well in hot dry lowland conditions but also up to about 2000 m.

SOIL TYPE AND CULTIVATION

The most suitable soils are well drained, loose, sandy soils. They will also grow on heavier soils, but this makes harvesting more difficult. Narrow ridges should be avoided and the seed beds must be prepared well.

Peanuts as nitrogen fixing plants are able to improve the quality of the soil substantially. They are therefore used as a rotation crop, because nitrogen which has been used by preceding crops, is returned to the soil.

SPACING, TIME OF PLANTING AND MATURITY

Usually one finger length between the plants and about three handspans between the rows is recommended for spacing.

The most suitable time for planting is the end of the wet season.

The time to maturity is from 3½ to 5 months, depending on the type of peanut. Relatively dry weather is required for ripening and harvesting. Peanuts are harvested when the top of the plant starts to die.

PESTS AND DISEASES

In many places peanuts are quite free from pests and diseases. In case of problems crop rotation is sufficient control.

Pineapples

ORIGIN

Pineapples originated in South America. It was not until the 19th century that pineapples were introduced to the Pacific region, including Australia and Papua New Guinea.

FOOD VALUE

Pineapples are high in sugar content and they also contain the vitamins A, B, and C.

Fig. 94 Pineapples *Ananas sativus* (Lindl.) Schult.

ALTITUDINAL LIMITS

Pineapples grow well up to an altitude of about 1700 m.

SOIL TYPE AND CULTIVATION

Well drained soils are best for growing pineapples. The addition of compost is essential, because pineapples require a lot of nutrients from the soil.

For example, when nitrogen is in short supply the leaves turn red.

Moderate to heavy rainfall is preferred for best growth.

Weed control by mulching and hand weeding is very often necessary. The land must be well cleared and should be dug about one handspan deep. The best place to prevent excessive weed growth and to protect the plants from too much direct sun is the shade of another crop, e.g. bananas. To prevent soil erosion, pineapples should be planted only on flat areas. If no other shade producing plant crop is available, dry grass or another dry material should be placed on top of the pineapple plants to protect them from too much sun (Fig. 95).

Fig. 95 Planting system of pineapples showing protection of the plants by dry grass scattered over the plants.

SPACING, TIME OF PLANTING AND MATURITY

Pineapples can be planted in rows or individually. One method is to plant two rows next to each other, about 3 handspans apart with one handspan between plants within the rows. A distance of one spade size in length between the center of the one double row and the next is recommended.

There are five different types of planting material that can be used:

Tops, slips, aerial suckers, ground suckers and butts.

Tops: The top or crown is the leafy part that grows on the top of the fruit. Plants that grow from tops take longer to fruit than those grown from suckers and slips.

Slips: These are the leafy branches that are on the fruit stalk just below the fruit.

Aerial suckers: These are leafy branches that are on the stem above ground level. Plants grown from aerial suckers produce fruit more quickly than other kinds of planting material, and yield more fruit.

Ground suckers: These are leafy branches that arise from the stem below ground level.

Butts: These are the stems on the plants that can be used for planting after the fruit is harvested. Plants grown from butts do not grow very well and are very slow to produce fruit. This method is therefore not recommended.

After a fruit has been harvested, the mother plant must be cut off and only one of the suckers left. The soil should be heaped around the base of the remaining sucker so that the roots can obtain food easily from the soil. This method is called 'ratooning'. It is not recommended to do this type of planting more than twice or in other words, to harvest more than three fruits from one plant, because the fruits become smaller with each ratoon harvest.

Therefore replanting of an aerial sucker is the best way to get a new, big crop.

It is best to plant the new crop on new ground and to plant a different crop on the old pineapple area. This helps to prevent the build up of weeds, insect, pests and diseases.

Planting can be done at any time of the year, provided the soil is not too dry.

Different planting material differs in the time taken for the fruit to be set and mature.

A crop grown from aerial suckers will produce fruit in about one year.

PESTS AND DISEASES

Pineapples do not have many pest and disease problems in Papua New Guinea. The mealy bug, a small white insect, or the wilt disease, caused by a fungus, can become problems.

Bananas

Musa balbisiana Colla

Fig. 96 Traditionally used banana variety (diploid variety)

ORIGIN

Edible bananas appear to have their origins in an area extending from Malaysia through Indonesia to New Britain in Papua New Guinea.

Musa paradisiaca L.

Fig. 97 Improved banana variety (triploid variety)

FOOD VALUE

Cooking bananas are an energy food, because of the starch content and ripe bananas are an energy and protective food because vitamins A and C are present in fairly high amounts.

ALTITUDINAL LIMITS

Bananas grow well up to 2000 m. They prefer humid areas.

SOIL TYPE AND CULTIVATION

Bananas can be grown on a wide range of soils, provided there is good drainage and adequate fertility and moisture. Bananas need alot of water and soil nutrients. Therefore compost and mulch greatly improves the growth conditions of bananas.

The land set aside for bananas should be cleared but not burnt, because the rotting material is a very suitable nutrient source for the plants. Holes must be dug for the plants at least one handspan deep and one handspan wide. A good supply of compost should be added to the hole before planting.

Any food crop, such as pumpkin or pineapples are suitable as a cover crop. The cover crop will prevent weeds from growing and in addition, they grow well in the shade of banana trees.

SPACING, TIME OF PLANTING AND MATURITY

It is best to plant bananas in rows on flat ground. Tall growing types are planted 1½ spade sizes apart, in rows 2½ spade sizes apart.

There are four types of planting material:

1. Bullheads or corms

2. Bits

3. Sword suckers

4. Maiden suckers

Bullheads: This is the underground part of the stem of a banana plant which has already borne a bunch. Because this is a very hard work to dig out this heavy part of the old banana it is not very popular.

Bits: This is a section of the bullhead. It must contain an eye before it can shoot. Bits are light, easy to transport and they grow well.

Sword sucker: Sword suckers have narrow sword shaped leaves. They are the most popular form of planting material. Suckers of ½ or 1 spade size in length are the best to plant.

Maiden sucker: These are taller former sword suckers. They are not easy to handle, but will produce fruits relatively early.

Suckers which come up on the mother plant before 6 months have passed, should be cut out. Any other ones should be removed, if the mother plant has already one sucker.

After the mother plant has produced one bunch, it is cut down and one sucker allowed to grow.

Planting is usually done at the beginning of the wet season, but it can be done all year round to provide a supply of bananas the entire year.

A bunch of fruit will normally be ready about 12–18 months after planting a sucker.

Planting material should be taken from plants that are known to bear well (see Figures 96 and 97 in comparison).

PESTS AND DISEASES

Flying foxes: They can cause damage to entire bunches of fruits. Covering the bunches with old banana leaves, sacks or other bags is the best way to prevent damage.

The best time to cover a bunch is after the last hand on the stem has been uncovered. The cover is tied to both ends, but make sure that a small hole is in the bottom to let out water.

Weevil borers: These are small insects that damage the banana corms. Therefore the suckers should be carefully chosen.

Nematodes: These are very small worms living in the ground which damage the roots of the plants.

Planting of cover crops will help reduce possible infection.

Leaf spot diseases: This kind of attack spoils the leaves, but there are varieties of bananas available which are resistant (e.g. Yava variety).

Papaya

Carica papaya L.

Fig. 98 Pawpaw showing the female (a) and male (b) flowers

ORIGIN

Probably originated from South Mexico and spread to all tropical and subtropical areas.

FOOD VALUE

The edible portion of the fresh fruit is very rich in vitamins A and C.

MEDICINAL VALUE

a. In case of constipation:
   For babies and children: Drink juice made from a ripe papaya. For adults: Eat a ripe papaya and in 1 day the constipation is cured.

b. Against worms in the intestine:
Make a paste out of the seeds of a ripe papaya fruit. Eat this paste together with other food (e.g. kaukau) for two days.

c. In case of purulent sore:
Place a piece of green, unripe papaya on top of the sore. Fasten it securely. Replace it every day.

d. For an open sore:
Place a clean leaf on top of the sore. Replace it every day.

e. To cure malaria:
Boil some leaves in water. Drink the cool broth. One cup for 5 consecutive days.

f. To cure a cough:
Boil cleaned roots in sugar water. Drink one cup of the broth for 3 consecutive days.

ALTITUDINAL LIMITS

It is found up to 1600 m in Papua New Guinea.

SOIL TYPE AND CULTIVATION

Pawpaw grows best in good deep topsoil with plenty of rotting matter in it. Wood ashes also encourage growth. Pawpaws are either male or female. Male and female trees can be recognized by their flowers (Fig. 98).

SPACING, TIME OF PLANTING AND MATURITY

Pawpaw is normally propagated from seeds. If sown in nursery beds, seeds should be about ½ a finger length deep at a spacing of one finger length. When planted, a spacing of 2 spade sizes by 2 spade sizes is sufficient. After flowering the sex of the trees can be determined.

Sowing can be carried out at any time of the year as long as adequate water is available.

Trees bear fruit at 12–18 months and they may live up to 25 years.

PESTS AND DISEASES

Fruit flies and flying foxes occasionally attack pawpaws, but generally the crop appears free from major pests and diseases.

Avocado

Fig. 99 Avocado fruit *Persea americana* Mill.

ORIGIN

Avocado originated in Central America and it was introduced in Papua New Guinea by Europeans.

FOOD VALUE

The fruit of avocado is rich in vitamin B and is a good source of vitamins A and E. The oil is highly digestible and makes a fruit of very high energy value.

ALTITUDINAL LIMITS

They can be grown up to 2200 m.

SOIL TYPE AND CULTIVATION

They will grow on most types of soil. Hand weeding may be done but deep digging may damage the shallow roots of the tree.

SPACING, TIME OF PLANTING AND MATURITY

Avocados can be easily grown from seeds but can also be grown from cuttings. The seeds are planted in the nursery two handspans apart in rows about 3 handspans apart. They can be placed on the surface of the soil or else covered by a very thin layer of soil. They can be transplanted to the garden area after about 12 months. Budding occurs after 2–4 months. Seeds can be planted at any time of the year. The transplanting can also be done at any time of the year.

Seedling avocados begin bearing after about 7 years, but vegetatively propagated ones usually bear earlier. Trees usually bear fruit every second year.

PESTS AND DISEASES

Root rot may kill the tree. Therefore trees should never be planted in poorly drained soil.

Guava

Fig. 100 Guava fruits *Psidium guyava* L.

ORIGIN

Guava originated in tropical America and it has spread throughout the tropics.

FOOD VALUE

Guavas are a particularly rich source of vitamin C. They contain about three times as much vitamin C as oranges and pawpaw. They are also a good source of vitamin A, iron, calcium and phosphorus.

ALTITUDINAL LIMITS

They grow from the lowlands up to about 1700 m.

SOIL TYPE AND CULTIVATION

No special cultivation is required and they grow in a wide variety of soils.

SPACING, TIME OF PLANTING AND MATURITY

Guavas can be grown from seeds or raised from cuttings by budding from known good producers.

The seeds are sown in nurseries to a depth of half a finger length. Germination will take 2–3 weeks and the seedlings can be transplanted when they are half a handspan tall.

Seedlings can be transplanted at any time of the year. Trees often bear their first fruits 2 years after transplanting. They will continue to produce for about 30 years, but they are at their peak after about 8 years.

PESTS AND DISEASES

Generally there are no insect pests or diseases of importance.

Citrus species

ORIGIN

Most citrus species like mandarin, lemon and lime, grape fruits and pomelos appear to have originated in China. Only

Fig. 101 Lemon fruits *Citrus limon* (L.) Burm.

during the last century the citrus species have been introduced to Papua New Guinea by Europeans. Mandarins and lemons are the hardiest of all cultivated citrus species and are grown today in most tropical and subtropical countries.

FOOD VALUE

The main food value in mandarins and lemons is the high amount of vitamin C. Also a moderate amount of vitamin A can be found.

ALTITUDINAL LIMITS

Lemons and mandarins grow well in Papua New Guinea on the coast and in the Highlands up to 2200 m. They grow best in an altitudinal range of 800 m to 1200 m.

SOIL TYPE AND CULTIVATION

Citrus trees can be grown on almost any type of soil, provided it is well drained. They prefer a light, loamy soil of good fertility and will not tolerate waterlogging.

For best results it is best to bud on to a locally adapted root stock. To produce root stock, seeds should be sown fresh, lightly covered in banana-fibre-pots, with sterilized soil. Germination occurs in 2–4 weeks. When the seedling has grown to a height of two handspans and the stem is about half a finger thick, it will be ready for budding.

Budwood should be taken from a tree with good fruits. The DPI staff will provide good budwood and show methods of budding.

The Liklik Book shows the budding process very clearly for all people who want to do it themselves. The budded seedlings will be ready for transplanting in about a year. The budded seedlings can be transplanted with the banana-fibre-pot straight into the cleared and prepared site.

SPACING, TIME OF PLANTING AND MATURITY

Planting distance will vary depending on the fertility of the soil, but about 5–6 spade sizes is an average spacing. The holes for the plants should be big enough that the roots are not crowded. Therefore the removal of the banana-fibre-pot is sometimes adequate, if the banana-fibre-pot is not yet rotting. Addition of compost to the planting hole supports the healthy growth of the young tree. Watering is necessary immediately after planting and mulching is very favourable.

Young trees should be planted out from the nursery at the beginning of the wet season. But any other time of the year is all right as long as enough water is available for the young tree.

Fruiting will start 3 to 4 years after being transplanted. The heaviest production begins about the fourth year after bearing.

PESTS AND DISEASES

A wide range of insects and mites attack citruses. Among the most serious are scales, aphids, fruit fly and leaf miner.

White wax scale is a thick waxy covering, common on leaves and branches of all citrus species. Both leaves and branches become covered with a sooty mould which grows on the honey dew excreted by the scales. No recommendation can be made to control this pest.

Aphids: These are small sucking insects which, when in large numbers, can damage plants through loss of sap.

Fruit Fly: The maggots or larvae of this insect feed on the ripening fruit.

Leaf miner: Larvae of these moths mine into the new leaves.

Root Rot: Trees suffering from root rot should be destroyed.

Soursack

ORIGIN

This fruit tree probably originates from tropical South America and has been introduced to Papua New Guinea by the Europeans.

FOOD VALUE

Its main value is the amount of vitamin C and A. It makes very nutritional drinks, ice cream and jelly.

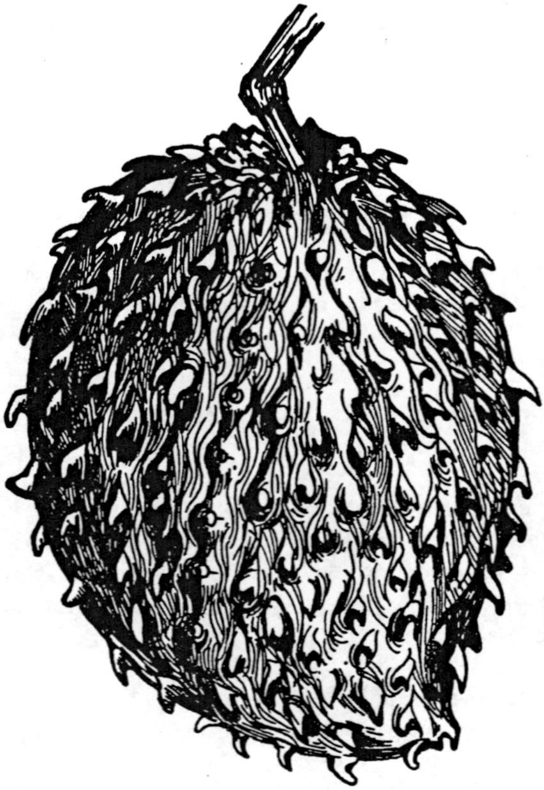

Fig. 102 Soursap fruit *Annona muricata* L.

ALTITUDINAL LIMITS

It will grow best up to 1000 m above sea level.

SOIL TYPE AND CULTIVATION

Any soil type rich in organic matter is suitable for growth. The tree can be grown from the seeds of the fruit. The seeds should be sown in banana-fibre-pots or in a nursery seed bed. The seedlings are ready to be transplanted when they are about one handspan high.

The trees should be pruned when they are about half a spade size above the ground. Only two shoots should be allowed to grow and these should be pruned again after a year. Repeated pruning will support a higher yield.

SPACING, TIME OF PLANTING AND MATURITY

Trees are usually planted 3 spade sizes apart. The fruits are usually large and dark-green and the surface is covered with numerous, recurved fleshy spines.

Any time of the year is good for transplanting but watering and a hole big enough so as not to squeeze the roots, prepared with some compost is required.

The trees fruit throughout the year but in order to assure a high yield of fruits, it is desirable to use hand pollination.

PESTS AND DISEASES

No serious pests and diseases are recorded in Papua New Guinea.

Mango

ORIGIN

It originates from tropical Asia and is regarded as one of the most delicious fruits of the tropics.

FOOD VALUE

It is a very good source of vitamin C and A. Fruit sugar and protein are also found in substantial amounts.

ALTITUDINAL LIMITS

It will grow well up to 500 m if enough rain is available.

SOIL TYPE AND CULTIVATION

Good fertile soil, rich in organic matter is most suitable for the optimum growth of the tree which can become very large.

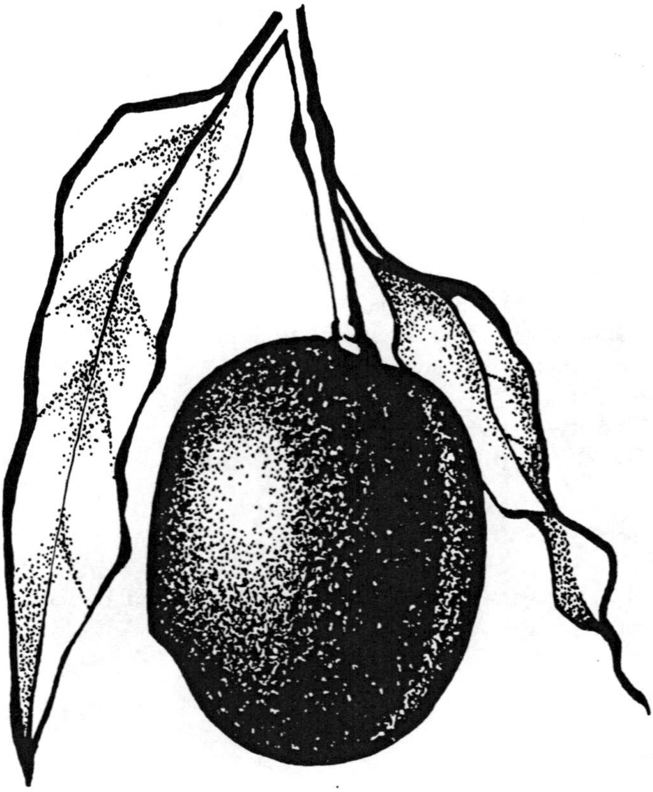

Fig. 103 Mango fruit *Mangifera indica* L.

The seeds will start to grow 8-14 days after being planted in a banana-fibre-pot or seed bed in a shaded nursery. The young plants are ready to be transplanted when the second lot of leaves start to grow.

Budding of the older plants will support the fruit production.

SPACING, TIME OF PLANTING AND MATURITY

The young trees are planted in holes which must be one spade size wide and half a spade size deep. The planting holes should be filled with compost and dark soil.

The trees should be planted at least 4–5 spade sizes apart.

Proper watering and care of the young trees is necessary.

Pruning is usually not necessary.

Any time of the year is suitable for transplanting, provided enough water is available for the young tree.

PESTS AND DISEASES

The larvae of the fruit fly can be a pest, but usually no serious pests and diseases are recorded in Papua New Guinea.

Pandanus

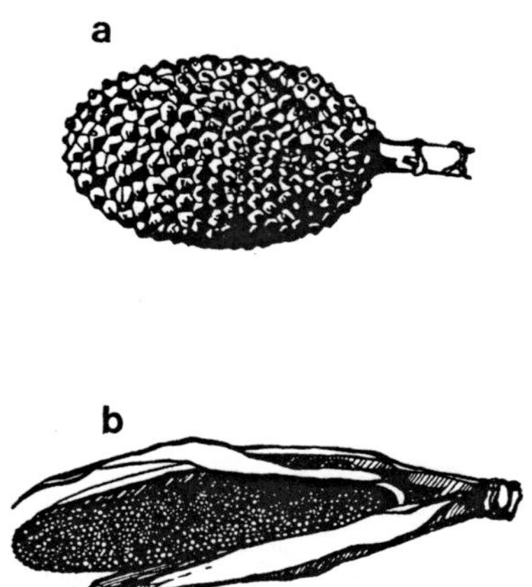

Pandanus tectorius Soland ex. Parkins.

Fig. 104 The two most popular pandanus species: Karuka (a) and Marita (b)

ORIGIN

Probably the more than 630 recorded species originate from different regions, like tropical Asia and America. In Papua New Guinea there are 66 different species.

FOOD VALUE

The fruit wall of the marita species is used to produce a dark red, oily sauce rich in vitamin A, fat and protein.

The karuka tree is a very important tree for the nuts which are rich in fat and protein.

ALTITUDINAL LIMITS

The marita species are chiefly lowland plants, but they can be cultivated up to 2000 m.

The karuka species are typically highland plants which do best above 2000 m. They can be grown up to 3000 m.

SOIL TYPE AND CULTIVATION

No specific soil type is required. The plants are generally cultivated vegetatively from cuttings.

SPACING, TIME OF PLANTING AND MATURITY

No specific spacing is required but it should be kept in mind that the plants will become large trees and will bear fruits as long as 50 years.

There is also no specific timing necessary for planting and usually karuka species fruit 5–6 years after planting.

PESTS AND DISEASES

Fungal and viral diseases cause "Yellow Spot" damage on the leaves.

Insects: Grasshoppers, larvae of butterflies and the "Banana Scab Moth" feed on the leaves and the floral parts of the plants. Wood boring beetles may cause extensive damage, leading to fruit fall and branch losses.

Passion Fruit

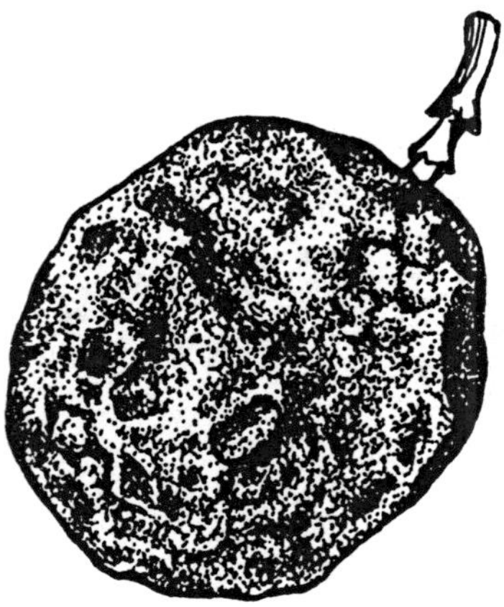

Fig. 105 Passion fruit *Passiflora edulis* Sims

ORIGIN

Passion fruits originate from South America and were introduced to Papua New Guinea by Europeans.

FOOD VALUE

The main food value of passion fruits lies in their substantial amount of vitamin C and A.

ALTITUDINAL LIMITS

The purple skinned passion fruit can be grown up to about 2400 m. The "banana" passion fruit will grow up to 3000 m.

SOIL TYPE AND CULTIVATION

Passion fruit can be grown in a wide variety of soils, but the soil should not be too wet.

The seeds must be planted first in a nursery or in banana-fibre-pots. The young plants should be raised in a shady place.

SPACING, TIME OF PLANTING AND MATURITY

Seeds should be planted in banana-fibre-pots or in a nursery in rows about one handspan apart. When the plants are about one finger high the plants can be transplanted in places where there is some kind of support for the climbing plants to grow. The direction of the growth of these climbing plants should be towards the sun (Fig. 106).

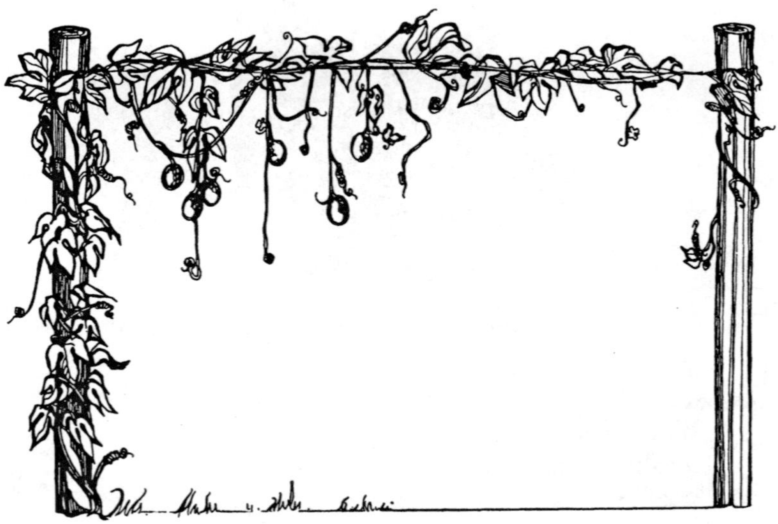

Fig. 106 . Arrangement for growing passion fruits

The plants can be transplanted at any time of the year. The first fruits may appear 12 to 18 months after planting. Fruiting is encouraged by pinching off the climber shoots or tendrils.

The main crop will take about 2 years to grow. When the fruit is ripe it falls to the ground and should be picked up as soon as possible after it has fallen.

PESTS AND DISEASES

The only disease that can become serious is "brown spot". Round brown spots appear on the leaves, stems and fruits. Plants or the part of the plant showing these symtoms should be burned.

Ginger

Fig. 107 Ginger plant *Zingiber officinale* Rosc.

ORIGIN

It originates from islands in the South Pacific Region.

FOOD VALUE

It is used as a spice for biscuits, cakes, tea, food and pickles, and as medicinal plant in some parts of PNG.

ALTITUDINAL LIMITS

It can be grown with good results up to 2000 m.

SOIL TYPE AND CULTIVATION

Loose, fertile soil is most suitable for the growth of this plant. The planting material should be carefully selected and should consist of small rhizome cuttings. The cuttings should be planted in small holes of one finger length in depth and should be covered with loose soil.

SPACING, TIME OF PLANTING AND MATURITY

The plant should be planted in ridges and the space between the individual plants should be about one handspan. Two handspans is appropriate for the space between the rows.

Planting is possible throughout the year and usually 8–9 months are required before the harvest of the rhizomes. The plants will then be about half or one spade size tall. The best time to harvest the rhizomes is when the shoots turn brown and begin to fall over.

PESTS AND DISEASES

Insect larvae sometimes bore through the leafy shoots. They are usually not a very serious pest and are often controlled by natural enemies like birds. Leaf spot and soft rot can cause serious damage. Careful selection of healthy cuttings and rotation are good preventive measures to avoid these diseases.

Chillies

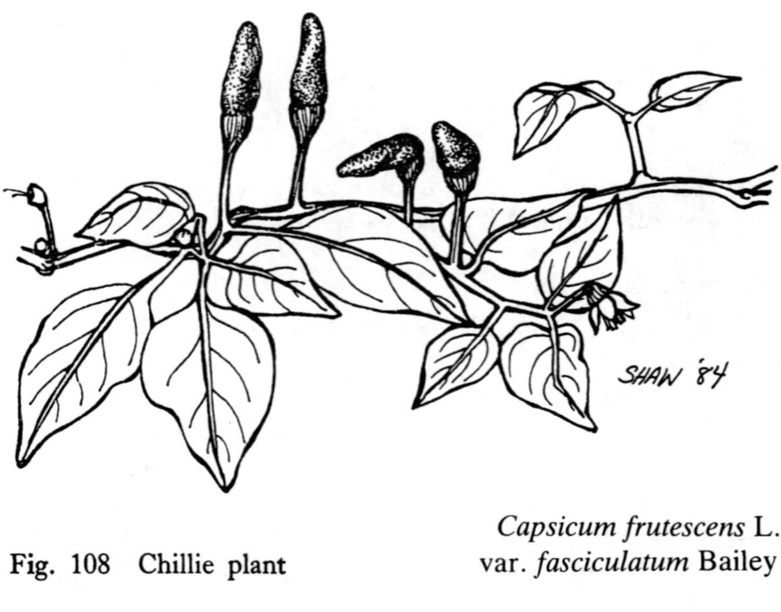

Fig. 108 Chillie plant — *Capsicum frutescens* L. var. *fasciculatum* Bailey

ORIGIN

Tropical and subtropical America is the origin of all varieties of peppers, including the chilli peppers which are usually about 9 cm long, narrow, pointed and very pungent. Chillies were introduced to Papua New Guinea only recently for the main purpose of selling them overseas.

FOOD VALUE

The main value in chillies is their pungency which makes them a valuable crop.

OTHER VALUES

Chillies are used in rotation or as border plants to help protect other vegetables from insect attack.

ALTITUDINAL LIMITS

Chillies will not grow in areas where there is frost. They grow best in places where there is some rain throughout the year and where the temperature stays below 30°C. Therefore the mid-mountain range between 800 and 2000 m is very suitable.

SOIL TYPE OF CULTIVATION

The soil should be soft and well drained. The best land for chillies is a flat or gently sloping land.

The land should be prepared the same way as for the food garden. Seedlings are grown first in larger banana-fibre-pots and transported to other parts of the garden once they have produced two true leaves.

The top of the young plants must be broken off about one week before transplanting. This makes the plants grow with more branches.

SPACING, TIME OF PLANTING AND MATURITY

The plants can be planted as border plants or in rows so that the person who picks the chillies can work easily between the rows of plants. One spade size should be the distance between the individual plants.

The best time for transplanting is on a rainy, cloudy day in the late afternoon. Proper watering of the young plants for about one week after transplanting is necessary.

About 4–6 weeks after transplanting the plants will start to flower. The fruit pods should be picked while they are green and thrown away. This helps the plant to grow more pods.

After 5–6 months the optimum number of fruits can be harvested and the plants will keep on producing fruit for 2–3 years. Only bright red and dry pods should be picked with their stalk. When the chillies are dried in the sun or solar drier, break off the stalk.

Warning: The juice of the pods and the dust of the pods can cause severe burning. Baking soda in water is the thing for washing and treatment of chilli burns.

PESTS AND DISEASES

Viruses, white mite and bacterial wilt can cause some damage to the plants. If the leaves are distorted, wilted or bronze coloured, one of these diseases has affected the plant.

Rotation is one of the best protective measures for these kinds of diseases. Agricultural officers can give some advice.

Information on subsistence agriculture improvement

Subsistence agriculture and soil conservation

Introduction

In nature all living things die and their death allows life to start again. When animals or plants die, their bodies rot and return to the soil as rich organic matter. This continual supply of organic matter helps plants grow and new life start to replace what has died, or been used. This is how nature maintains a balance between life and death.

Gardening should follow nature's balance. Leaves, grass, animal manure and plants should be returned to the soil and used again, not thrown away (Fig. 109).

Fig. 109 Natural soil layers in a forest area before its use as a garden
1 Fertile forest floor, 2 Fresh organic material, 3 Breaking down materials, 4 Topsoil, 5 Fine rock particles, 6 Larger rock particles, 7 Rocks

This system is in balance if traditional subsistence agriculture systems of cut and burn or cut and rot are used, but only under the following circumstances:

— Plenty of land and forests available
— Only a few people in a given area
— Plenty of animals and edible subsistences available
— No high technology employed

As a result of a population increase, and the development and introduction of improved technology, like axes, knives, chain saws, bulldozers, guns etc., the forests are shrinking rapidly and many plants and animals are disappearing. Nature's balance is very often seriously disturbed. Grasslands and wastelands are increasing and arable land for food production is decreasing rapidly in both size and quality.

Under these changing conditions we also have to change if we do not want our life situation to become worse. Therefore some parts of the traditional subsistence agriculture practises have to be changed and replaced with new practises or else supplemented with a range of more promising methods which are acceptable to the subsistence farmer and which improve the life situation of these rural people.

Shifting cultivation

Shifting cultivation is an agricultural system whereby people cut and burn or cut and rot a given jungle area, farm this area for about 1–3 years and then abandon it afterwards and move to a new garden area in the forest, usually returning to the old garden site in about 10–15 years. By this they have formed a circle around the village (Fig. 110).

As the population of an area increases, the length of the fallow period shortens, therefore the existing land is over-used which leads to a steady decrease in food production if no improvement measures are employed. If some of the fertile land is used for cash crops, like coffee, even more pressure is put on the area used for annual subsistence crops.

When this happens, the traditional subsistence agriculture system is no longer sufficient to produce all the necessary food. Therefore, we have to find ways to improve the subsistence farming in a given area.

Alternatives to shifting cultivation

Some countries in tropical areas, which have similar conditions to Papua New Guinea, have already been forced to devise systems of permanent farming or site stable gardening because

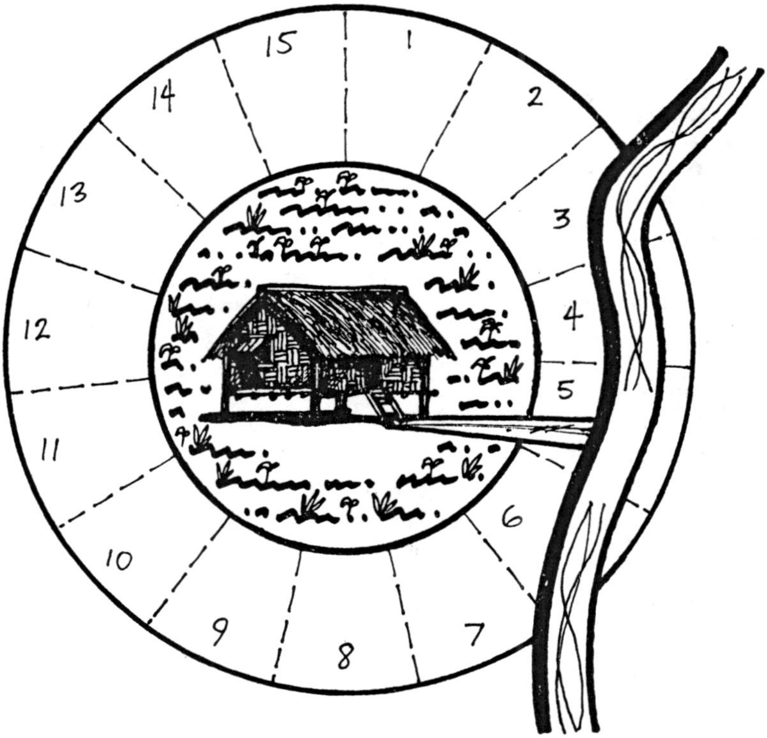

Fig. 110 Shifting cultivation pattern around a village
1—15 years

of the problems mentioned above. The type of system that has been developed usually involves one or more of the following methods which suit their way of life.

These methods can be listed as follows:

— Cultivation of rice in irrigation areas to produce enough stable food for a given population
— Substituting grazing pasture to pioneer bush or grass fallow between cropping periods
— Increased use of animal and even human manure and the extensive use of compost
— Rotation of crops and the use of crop hygiene methods to control pests and diseases
— Use of fertilizers, and pesticides to reduce pests and diseases

— Use of nitrifying plants as rotation crops or green manure to improve the soil nitrogen level
— Improvement of soil conservation measures
— Establishment of agroforestry methods

Reasons for soil erosion

Most soil erosion is caused by man, often because he practises a type of agriculture not suited to the type of land he is cultivating. Soil erosion takes place in many different ways and at different rates. It often depends on one or the other of the following factors:

— topography of the area
— amount of rainfall
— intensity of rainfall
— type of crop grown, e.g. whether they cover the ground or whether they are deep rooting
— depth and quality of topsoil

Soil conservation techniques

There are various useful techniques being utilized to conserve the soil. Two main categories of soil conservation techniques are:

1. Biological conservation measures: This includes the use of living or dead biological material to minimize erosion processes.

2. Physical conservation measures: This includes methods of land shaping to contain and control run-off water and by this means prevent soil erosion.

1. BIOLOGICAL CONSERVATION MEASURES

a. *Agronomic measures*

These measures favour the rapid establishment of a dense ground cover suitable for the subsistence farmer. They include the following methods:
— Compost application
— Green manure crops
— Crop rotation
— Mixed cropping
— Planting of crops at first rainfall, having prepared the land during the dry season (Fig. 111)

Fig. 111 Possible events in a shifting cultivation area
a Undisturbed forest area
b Cut and burn procedure of shifting cultivation
c If the soil is left bare the rain will soon wash the nutrients out of the soil

d

e

f

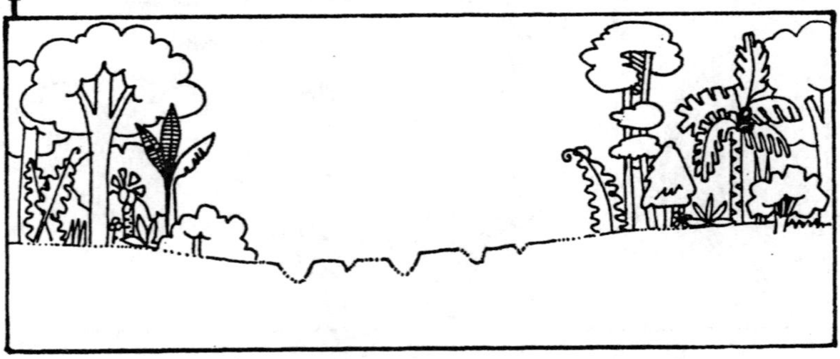

d If a crop is grown immediately after the clearing process the crop plants will grow well and protect the soil from being washed out

e If the old crop is dug into the soil the decomposed organic material will provide nutrients for the next crop

f If the old plants are totally removed or burned, the rain will wash away the soil and ruining the garden

b. *Mulching (Fig. 112)*

Fig. 112 Mulching practices in a garden with leaves and tree litter

Mulching is a method used to cover the soil with organic material such as leaves, grass or tree branches. Mulching favours the following effects:

- Reduction of the size of the raindrops reaching the base of the plant
- Cushioning the raindrops at the soil surface, preventing the break down of the soil and the splashing away of the soil and thus reducing of the compactness of the soil
- Storage of the soil moisture and cooling the soil during the dry season
- Keeping the weeds down

— Extensiou of the growing season and therefore extension of the period of effective plant ground cover

c. *Minimum or No-Tilage*

This is a type of crop cultivation system whereby a minimum amount of ground is disturbed in ground preparation necessary to plant a crop. For example, the traditional method of planting taro, banana or other crops using only a digging or planting stick is such a system.

d. *Contour planting*

This is the planting of a crop in rows along the contour lines of an area. This type of planting arrangement can be done with one crop or a mixture of crops, annual or perennial crops or intercropping of annual and perennial crops.

e. *Strip cropping (Fig. 113)*

Strip cropping means that alternate strips of grass or bush and food crops are laid in contours. The grass or bush cover provides barriers to run-off water which on cropping a bush strip will likely slow down its speed and therefore reduce its capacity to carry away soil. After a number of years bush strips can be cultivated and the cultivated areas are then allowed to go back to bush again.

Modifications of this system can be developed to suit various parts of Papua New Guinea. Instead of grass, fast growing nitrogen fixing trees or other useful tree crops can be planted. This is a kind of shifting cultivation with a very much reduced fallow period.

Fig. 113 Strip cropping in a forest area

f. *Rotation (Fig. 114)*

Fig. 114 One example of rotation of garden crops
a First crop: Corn, b Second crop: Peanuts, c Third crop: Taro

This is somewhat similar to a short fallow system. It is well known that the soil depletion from annual crops is much faster than with perennial crops. Therefore it is necessary to rotate annual and perennial crops and balance the losses in nutrients and soil formation. Another very important effect of rotation is the prevention of pest and disease build up in a given garden area.

2. PHYSICAL CONSERVATION MEASURES

a. In some areas logs are placed in the garden area so that they prevent soil erosion (Fig. 115). This is a traditional form of an anti-erosion method used in the highlands of Papua New Guinea.

Fig. 115 Traditional soil conservation method by placing logs in the garden area

b. Contour ridging is a method of planting crops such as corn, peanuts or kaukau (sweet potato) along the contour line of a gentle slope.

c. Contour beds or terracing of a sloping area for broader garden beds (see Fig. 58), and on steeper slopes.

d. Individual terracing is especially advisable for tree crops (see Fig. 68).

e. Drainage system of a given garden area to get rid of excess water.

f. Setting of silt traps as a sort of reverse drainage to trap all soil or silt washed away by the run-off water. These silt traps should be dug in the drainages or the paths between the garden beds. They can also be used as dropping places for weeds, which will decompose, adding nutrients to the silt. The silt should be transferred back to the garden beds from time to time (Fig. 116).

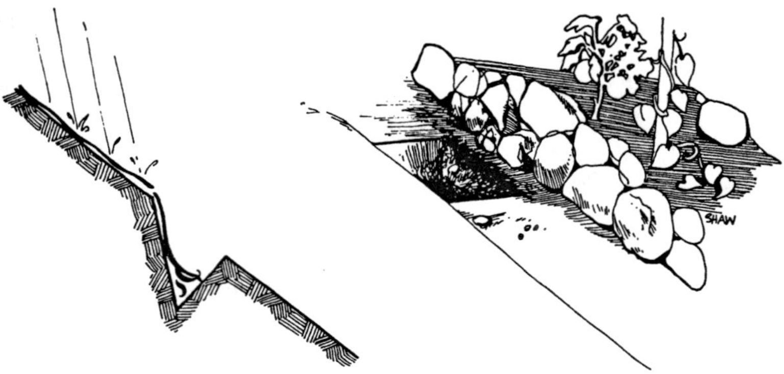

Fig. 116 Silt traps in the walking space between the single garden beds

3. OTHER PREVENTIVE MEASURES

Besides the two mentioned conservation measures there are some other measures which help prevent soil losses. Many of the causes of soil erosion can be prevented by:

— Avoiding the burning of leaves, stems, bushes or any area during dry season. This causes the soil to lie bare and exposed to heavy rains and causes much of the nitrogen, sulphur and carbon to be lost in the smoke, instead of being used for mulching or for the production of compost.

— Avoiding cultivation on too steep hillsides.
— Avoiding cutting all trees in a given garden area.
— Leaving the leaves and other organic matter on the surface of the garden area.
— Avoiding gardening too close to river banks, creeks or edges of landslides. A distance of at least 50 meters or 30 spade sizes should be between the garden and the affected area.
— Avoiding cultivating or tilling too much of the area at one time, especially in very hilly areas.
— Avoid completely cleaning away all weeds in drains and along the roads in the garden area.

Conclusion

Traditional shifting cultivation with long fallow periods is regarded as one of the best and most appropriate methods of subsistence farming, as long as it is not subject to population pressure. An increase in population will rapidly change all the beneficial aspects. With fallow periods getting shorter and shorter soil degradation is accelerated tremendously.

Therefore, the villagers have to be taught the concept of soil conservation by practising with them appropriate techniques in their garden areas.

Site-stable gardening could be an alternative to the prevent shifting cultivation systems. If accompanied by appropriate conservation techniques and soil improvement techniques the site-stable gardening has good prospects.

Agroforestry – a short overview

Introduction

Traditional subsistence agriculture in Papua New Guinea involves:

— Clearing of forest lands by slash and burn or slash and rot measures.
— Frequent rotation of garden sites.

This practise is said to be ecologically well balanced where land is plentiful and the population is low. However, some areas in Papua New Guinea are facing increasing land pressures for subsistence food production as a result of:

— increasing population.
— transfer of valuable land to cash cropping.
— other uses of land.

The garden fallow periods have been critically shortened. Traditionally, a long fallow system has been used (Fig. 117). But in most of the mid-mountain areas a short fallow period is now practised (Fig. 118).

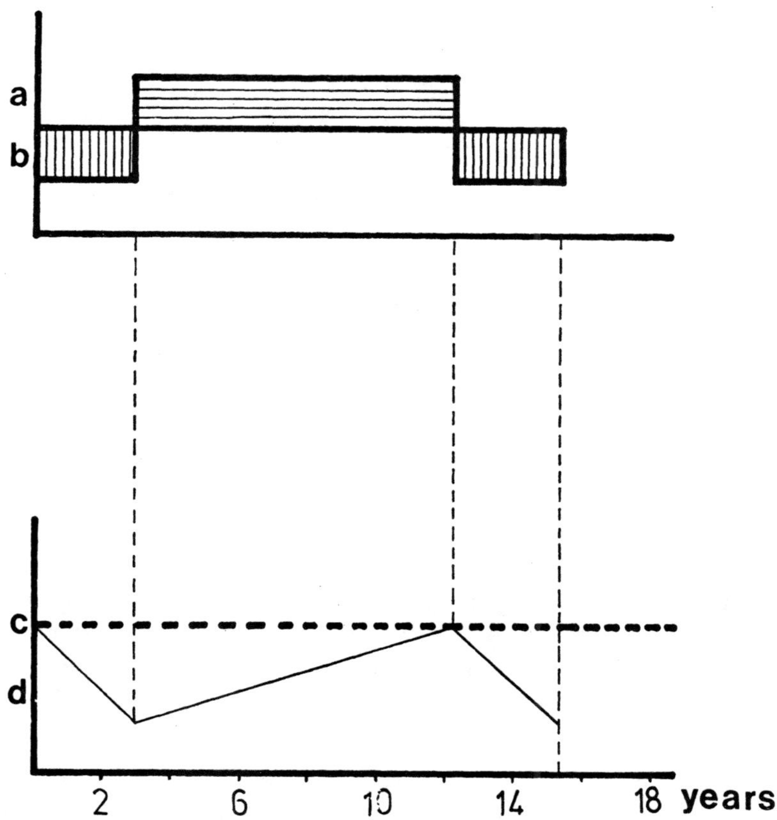

Fig. 117 Diagram giving the system of long fallow
a Fallow, b Cropping, c Soil: Original nutrient level, d Reduction of nutrient level

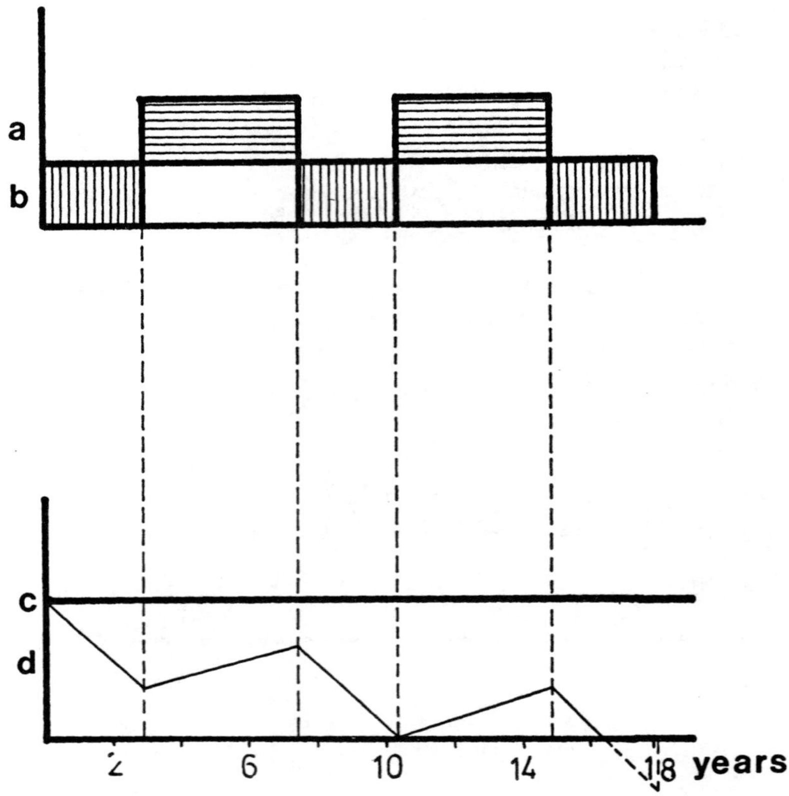

Fig. 118 Diagram giving the system of short fallow
a Fallow, b Cropping, c Soil: Original nutrient level,
d Reduction of soil nutrient level

The shortening of the fallow period results in the establishment of permanent grass lands, a decline in soil fertility and biomass of regrowth, an increase in soil erosion and a shortage of firewood and wood for construction purposes.

Definition of agroforestry

Agroforestry is a type of land use system where forestry trees are combined with food crops, or livestock or a combination of both, on the same unit of land either simultaneously or sequentially, with the aim of improving or maintaining the production in a sustainable manner.

Agroforestry concept

The concept of agroforestry is combining tree planting with garden production to control soil erosion, improve the soil conditions, useful wood products and maintain high production of food crops and other materials from the same unit of land indefinitely.

Component combination

This is the combination of the three main agroforestry components, trees, food crops and livestock, in different ways to come up with various combinations of cropping systems (Fig. 119).

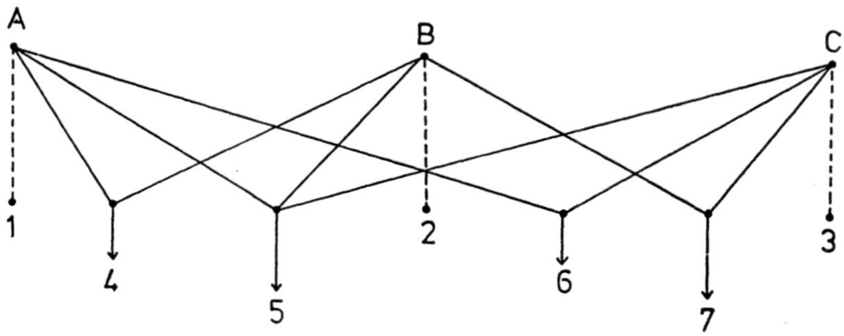

Fig. 119 Agroforestry system based within component combinations

Components	Combinations
A Trees	1 Forestry
B Food Crops	2 Agriculture
C Livestock	3 Ranching
	4 Agroforestry in strict sense
	5 Agro-silvo-pastoral System
	6 Silvo-pastoral System
	7 Agro-pastoral System

1–3 Non-agroforestry combinations

5–6 Agroforestry in wider sense

This includes the following main systems:

1. Shifting cultivation or cyclic cultivation system (Fig. 120)

This system alternates food crops with forest fallows.

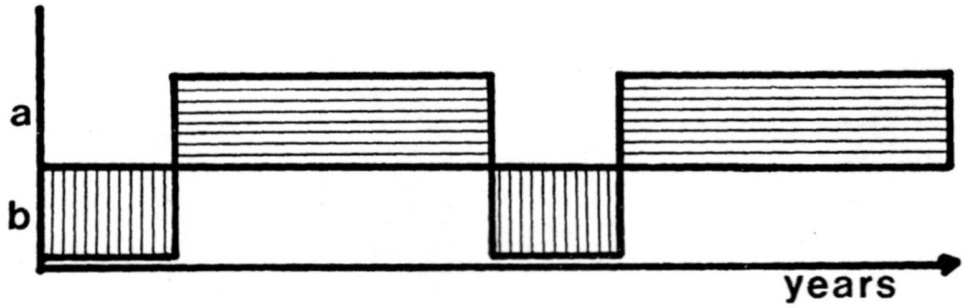

Fig. 120 Cyclic cultivation system of shifting cultivation (after Vegara, 1982)
a Fallow b Cultivation

2. Taungya System (Fig. 121)

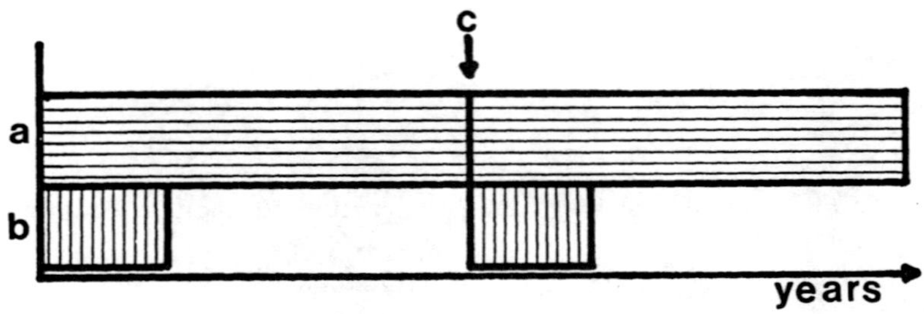

Fig. 121 Taungya System of shifting cultivation (after Vegara, 1982)
a Forest plantation, b Cultivation, c Harvest and regeneration of the forest plantation

Food crops and tree crops are planted simultaneously. After a few years the food crops cease due to the shadowing effects of the grown trees. The forest is allowed to grow for about 10 years.

Then the forest is harvested for timber and firewood and afterwards the area is again re-established with crops and trees simultaneously.

3. Integral or simultaneous agroforestry (Fig. 122)

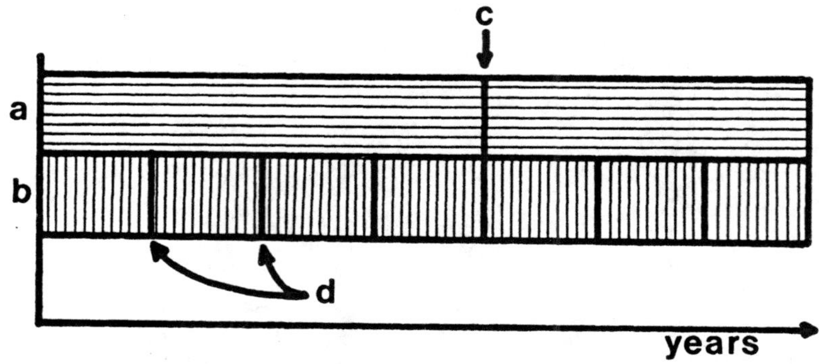

Fig. 122 Integral or simultaneous agroforestry (after Vegera, 1982)
a Forest Crop, b Cultivation, c Harvest and regeneration of tree crops, d Harvest and replanting of food crops

Food crops and trees are planted simultaneously on the given area. The spacing and garden arrangements are set up in a way that the food crops can be planted and harvested continuously. After about 10 years the trees are harvested as timber or firewood.

Food crops in agroforestry

Food crops for home consumption will be the most important product in agroforestry farms for the subsistence farmers. These food crops are mostly annual ones rather than perennial ones. The need for continuous food production and the farmers' desire for a permanent land tenure suggests that the integral or simultaneous agroforestry system is the most suitable one for the subsistence farmers' needs. Of great importance in this system is the space arrangements between the different crops of trees and food crops. There are four types of space arrangements in use:

— Boundary planting of border trees (Fig. 123)
— The alternate row planting system (Fig. 124)
— The alternate strip planting system or alley cropping system (Fig. 125)
— The random planting system (Fig. 126)

Fig. 123 Boundary planting of border trees (after Vegera, 1982)

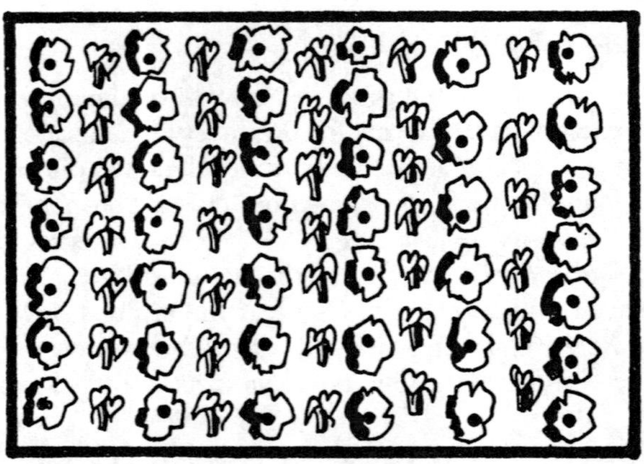

Fig. 124 The alternate row planting system (after Vegera, 1982)

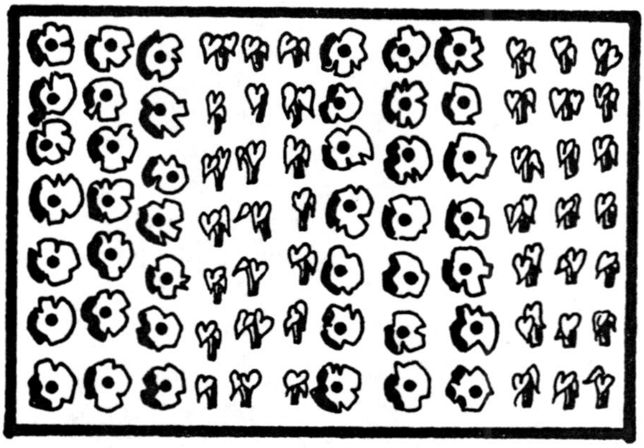

Fig. 125 The alternate strip planting system or alley cropping system (after Vegera, 1982)

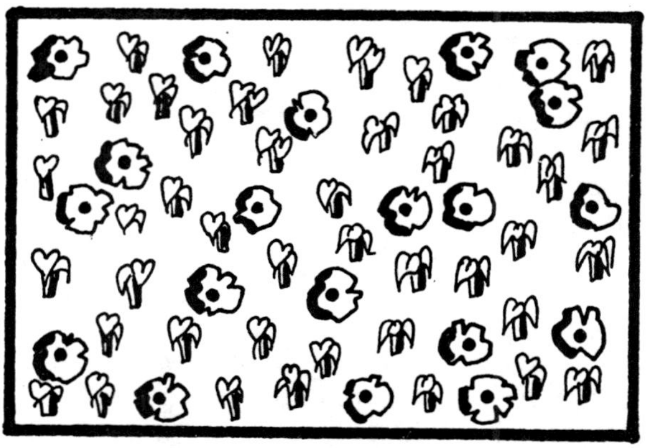

Fig. 126 The random planting system (after Vegera, 1982)

The division of land used for food and tree crops should be 50–90% in favour of the food crops. For the improvement of the soil, nitrifying trees should be integrated in a certain amount in the tree planting process.

Benefits from agroforestry

The trees integrated in this system can provide the following benefits:

— Adding nitrogen to the soil if these trees are nitrifying trees like *Leucaena* species.
— Protection of the area against soil erosion by fixing the ground with their rooting system, and with leave litter.
— Adding valuable organic matter by distributing leaf litter.
— Maintaining the soil moisture by shadowing the area and covering it with leaf litter as a sort of mulching.
— Increasing the diversity of different plants on a given area and by this means contributing to fewer insect pest damages.
— Reduction of weed growth by shadowing and leaf litter contribution.

However, there are also some detrimental effects like the reduction of sunlight, competition of space and nutrients with the crop plants, favouring some fungal diseases, increasing the moisture of the soil to an unfavourable level and even inhibiting the growth of some crop plants.

Multiple advantages of agroforestry

1. ECOLOGICAL ADVANTAGES

It is of great importance in nature that the diversity of species, whether plant or animal species is as great as possible in a given area. Agroforestry methods do increase diversity.

Besides the advantages of diversity increases, the even distribution of sunlight is of great importance for the balanced development of ecological units in nature. The deep rooting of most of the trees will certainly reduce or prevent soil erosion and if specific trees are planted the improvement of the soil fertility will not only be beneficial for the crop plants, but also for all the other plants in this area.

2. ECONOMIC ADVANTAGES

By using the agroforestry system there will result a greater overall diversified production of goods for the subsistence farmer, giving him an alternative income and a regular one throughout the year. It seems that agroforestry is suited to satisfy the necessities of the subsistence farmer than any other farming method in Papua New Guinea.

3. SOCIAL ADVANTAGES

The most important factor is the relevancy of agroforestry to the culture of the subsistence farmers. It certainly will improve his nutrition and will create a greater self-sufficiency. This system will not interrupt or completely change the life style of the rural people but it will improve the problems arising from the steady build up of the population pressure.

Nitrogen fixing trees

Introduction

Nitrogen fixing trees are considered crucial to the balance of nature, as many are able to convert nitrogen gas from the air into ammonia which is a soluble form of nitrogen utilized by other associated plant communities. The nitrogen input of nitrogen fixing trees can be extremely important for maintaining soil productivity over long periods. A nitrogen fixing tree crop has been recorded to contribute up to 500 kg of nitrogen to the soil per hectare per year.

With increasing emphasis on the introduction of agroforestry to rural and even urban communities, nitrogen fixing trees are considered to have a vitally important role to play in agroforestry systems because of their potential to fix nitrogen and for the other products they can provide.

Nitrogen fixing tree genera

The ability to fix nitrogen has often been attributed to the legume family which contains vines, shrubs and trees. However, other plant families have been shown to have nitrogen fixing abilities, and in time other families may be demonstrated to have this ability.

Some tree families that have plants with nitrogen fixing abilities are:

Family Name	Genera
Casuarinaceae	Casuarina, Gymnostoma (Fig. 127)
Betulaceae	Alnus
Leguminosae	Acacia, Albizia (Fig. 128), Leucaena (Fig. 129) Sesbania
Ulmaceae	Trema, Paraspondia

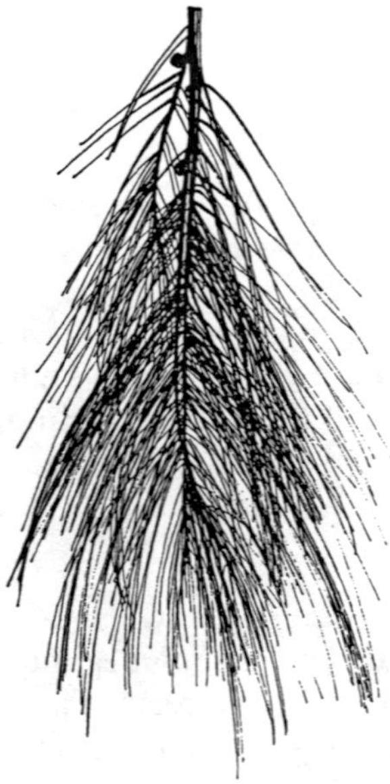

Fig. 127 The Yar tree *(Casuarina oligodon)*

Of the four families, the Leguminosae have by far the largest number of tree genera and species with some potential for inclusion in agroforestry and possibly offer a greater number of uses.

Other roles of nitrogen fixing trees

Many nitrogen fixing trees can be used for a variety of end products. A most common and valuable end use is for fuel wood either in log or charcoal form. Examples of this may be found with *Leucaena* and *Casuarina oligodon,* the Highlands "Yar"

Fig. 128 *Albizia falcataria*

tree. Some genera and species produce valuable construction and veneer wood. In the case of the latter an excellent example is *Pterocarpus indicus* known as the "Rosewood" tree. Other genera and species can be used as a food source. A number of Leguminous trees produce edible beans and seeds and their foliage, in some cases, may be used to feed livestock. Finally, many nitrogen fixing trees are extremely useful for shade and for erosion control.

Research with nitrogen fixing trees in Papua New Guinea

Research with nitrogen fixing trees in Papua New Guinea is comparatively recent although the use of trees such as *Casuarina oligodon, Albizia chinensis* and *Leucaena leucocephala,* have been in use as nurse crop trees for many years.

Fig. 129 *Leucaena leucocephala*

Current research in Papua New Guinea is involved with species selection and introduction for inclusion in agroforestry and forestry projects. Some research into management techniques has also been implemented with some giant strains and species of *Leucaena*.

The selection of a nitrogen fixing tree species depends on a number of considerations:

1. ADAPTABILITY

Selection will depend on the tree species ability to live on the sites it is intended for. A tree species should be tested over a range of sites and altitudes.

2. COMPATABILITY

A species must be compatible with other plant species that will be grown with it.

3. VERSATILITY

Although not entirely essential, a tree species that can be used for many purposes is of great value in an agroforestry system. Primary traits that are considered are growth rate, stem straightness and length of utilizable bole for fuel and lumber. The straighter the tree the more end uses it is likely to have. Food and fodder production, the ability to reproduce vegetatively or by seed are important considerations.

4. ADOPTABILITY

It is important when considering a tree for selection that it does have some use and that it will be of some value to those for whom it is intended and that they also understand how it is to be used. Also the weed potential of a tree species must be taken into account as every step must be taken to insure that a species will not become an unmanageable pest.

Once species of value have been identified, future research will most likely look at ways of improving the tree productivity and use through selecting and breeding and making this seed available to the general public and to industry.

Nitrogen fixing tree species in forestry trials

The following are tree species that are being observed and measured in trial plantings and proper experiments. Some comments about these are provided. It will be noted that these are all leguminous trees.

Albizia falcataria

A fast growing tree, native to Papua New Guinea and parts of Indonesia. Popular in South East Asia as a species for reforestation and afforestation projects. The species grows into a very large tree and produces a light, distinct flat crown. It has been considered a useful nurse crop for *Araucaria,* established as underplantings and because of its light shade will permit pasture development and certain forms of cash and subsistence under cropping. Its main disadvantage is size and branch shedding habit. The species is compatible with other native tree species that are relatively shade tolerant themselves.

Albizia chinensis

Also a very fast growing tree introduced from South East Asia and other countries for shade in coffee plantations. It produces denser shade than the previous species and coppices well. Although it can be used as a timber tree it has no potential in Papua New Guinea forestry. It can be used as a nurse crop for other trees such as *Araucaria* (Fig. 130), *Agathis* (Fig. 131), *Canarium* (Galip) (Fig. 132) and *Terminalia* (Fig. 133). It may have limited use in agroforestry because of its dense shade. However, the species does shed most of its leaves at various periods which may have some advantages in certain undercropping systems.

Fig. 130 *Araucaria klinkii*

Acacia species

Rather than deal with individual species the four native *Acacia* species: *A. auriculiformis* (Fig. 134), *A. aulacocarpa*, *A. crassicarpa* (Fig. 135), *A. mangium* (Fig. 136), are only just

Fig. 131 *Agathia alba*

Fig. 132 *Canarium sp.* (Galip)

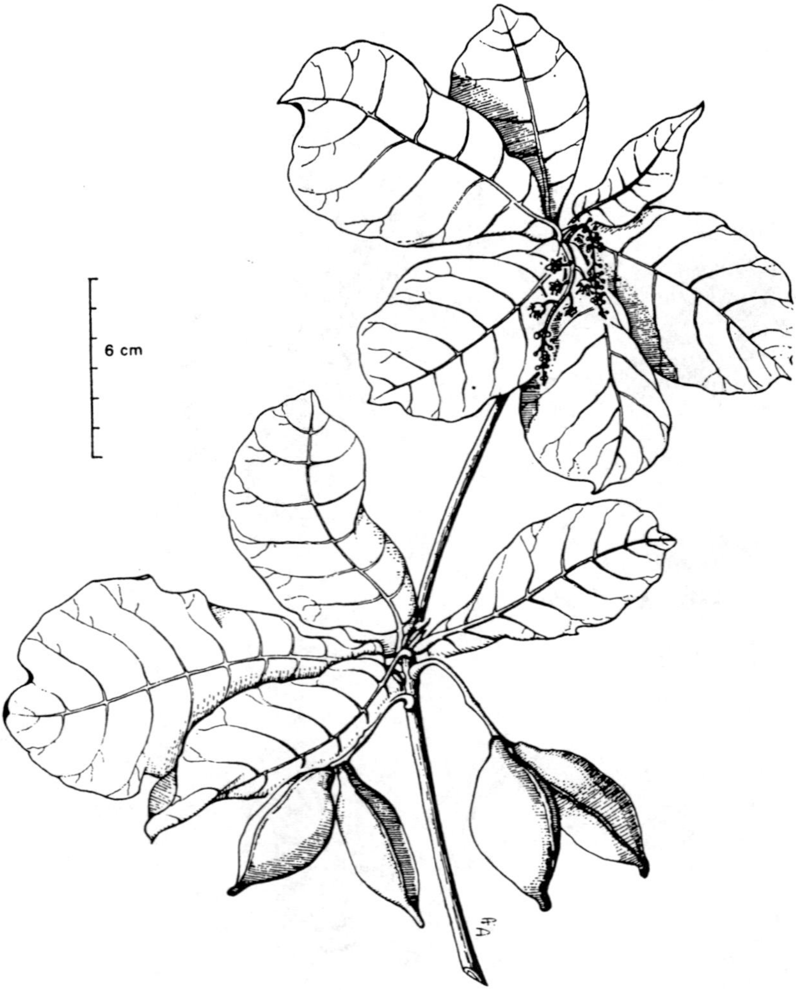

Fig. 133 *Terminalia catappa*

Fig. 134 *Acacia auriculiformis*

being tested in forestry in Papua New Guinea to assess their potential and to look for significant and useful differences in specific traits between natural populations. At present it is felt these have limited use in agroforestry except for shade, wind protection, fuel wood, wood chip, lumber and tannins.

Calliandra species

Some five species have been introduced into Papua New Guinea, mainly for ornamentals. However two apparently recent introductions, *Calliandra calothyrsus* (Fig. 137) and *Calliandra houstoniana,* are reputed to have considerable value as a fuel wood, shade and fodder species. They are also useful for honey production. These plants grow into low spreading trees and shrubs and because of their size would be easily managed. Further research is required to assess the species potential in forestry and agroforestry situations.

These South American species are being commonly used in Indonesia in village agroforestry plots and in larger plantations for fuel and fodder indicating the species potential for Papua New Guinea agroforestry.

Fig. 135 *Acacia crassicarpa*

A — Habitus   *B — Seeds*   *C — Flowers*

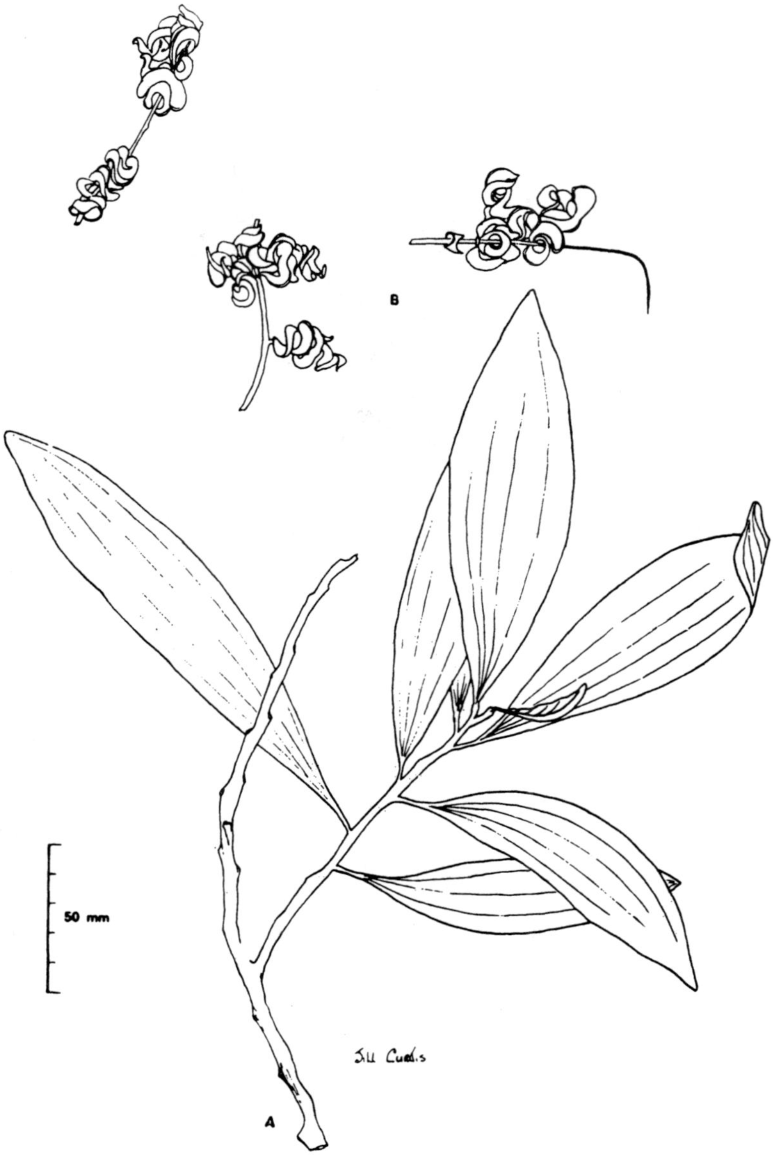

Fig. 136 *Acacia mangium*

A — Habitus *B — Seeds*

Fig. 137 *Calliandra sp.*

Leucaena species

The common *Leucaena leucocephala* (See Fig. 129) is wide spread throughout the Pacific region. It is commonly used as a nurse crop for cocoa and coffee in Papua New Guinea and is utilized as fodder and for light building material and fuel wood. There are at least ten species of *Leucaena,* found in parts of South America, and a number of these, grow to a much greater size than the wide spread low growing form of *L. leucocephala.* The Office of Forests initially became interested in giant forms of *L. leucocephala* because of their fast growth and wood chip potential. In the course of researching giant *Leucaena,* five species and a large number of giant strains of *L. leucocephala* have been introduced for trial and seed production during the past six years. The species introduced are: *L. collininsii, L. diversifolia,* L. sp. ex Mexico, *L. leucocephala* (Fig. 138) and *L. pulverulenta.*

Considerable differences have been noted in growth and habit between species and their varieties and from these trials it has been possible to identify species suitable for specific areas

Fig. 138 *Leucaena leucocephala* ex Mexico

and select and recommend specific end uses of these. Among the best performers of note so far, is the giant mexican species. It is well suited to planting sites, with soils of moderate fertility to low fertility, at altitudes from 700 to 1500 metres or possibly higher. The species has specific characteristics which make it attractive for agroforestry. They are fast growth over a 12 month period, good, erect stem form, making it suitable for construction work, and high wood density for fuel wood and light crown producing adequate light for undercropping. It is expected that further research will be conducted with other species of *Leucaena* as they are considered to have further potential for fuel, food forage, lumber and for site amelioration in agroforestry. Future research work will continue to identify species and strains suited to specific altitudes and soils and carry out selection and tree improvement work.

Casuarina – a nitrogen fixing tree

No paper on nitrogen fixing trees would be complete without mention of the *Casuarina* species. Most Papua New Guineans, who live in or have visited the Highlands, are familiar with the widely planted "Yar" tree of *Casuarina oligodon* (See Fig. 127). The tree is used for coffee shade and as a source of building material and fuel wood. The species is recognized as putting "grease" back into the soil or, in other words, increases the soil fertility through the addition of nitrogen and leaf litter.

The genus *Casuarina* has recently undergone a taxonomic revision with a second genus *Gymnostoma* now established. In this paper they are collectively refered to as *Casuarina,* and together there are around 98 species, many of which are considered to have considerable potential in agroforestry, for erosion control, for fuel and fodder and for lumber. A number have already demonstrated their ability to adapt to different ecological sites in a number of countries.

Casuarina oligodon is mainly a mountain species and does not thrive in the lowlands, but another natural occurring species, *Casuarina equisetifolia,* does occur on the coast or nearby. This species has yet to be fully tested as it appears to have considerable potential in the lowlands for fuel wood plantations, and as a nurse crop tree in agriculture. Other exotic species have not been tested as far as is known, although some have been introduced as ornamentals. It is felt that there may be considerable scope for *Casuarina* research in Papua New Guinea, both with native and exotic species. However, if contemplating such research, one must be careful that no diseases of the *Casuarina* are introduced with seed imports, as an exotic attack among the "Yar" would have serious consequences for the rural population.

Establishing nitrogen fixing trees

THE NURSERY PHASE

Nitrogen fixing trees may be raised by seed or, depending on the nature of the species, by cutting. In most cases the plants are likely to be raised by seed.

In forestry nurseries, seeds of such species are generally directly sown into the tubes of soil and arranged in rows in wire support stand out beds. In some situations they may be shaded until the seedlings have all emerged. The plants are usually planted out when they are around 40 cm high.

When the species to be planted is a legume, there may be a need to treat the seeds by immersing them in boiling water for up to 4 minutes, then soaking the seeds in cold water for 24 hours, to speed up and increase germination. Alternatively the seed coat may be cut with an abrasive, such as sand paper, to induce speedy germination. In some cases seeds of legumes may require coating with inoculum to produce nitrogen fixing rhizobia nodules on the roots. Your local agricultural officer should be contacted on this for advice and assistance in acquiring the correct inoculum and how to apply it to the seed.

THE PLANTING PHASE

The spacing at which the trees will be planted will depend on the growth habit of the species and the purpose for which they are planted. For fuel production and inter-cropping with *Leucaena* for example, the distance between rows may range from 3 to 4 metres, or even up to 8 metres. This will depend to some extent on the amount of shade the particular species of *Leucaena* will provide. The distance between trees in the rows can range from 1 to 3 metres which will provide ample material to be harvested by mechanical thinning, on an alternate tree and alternate row basis. Planting should be done with only the healthiest trees and preferably those of even size. The planting should naturally be done during the local wet season when the soil is moist and will remain so for a sufficient period for the seedlings to establish.

MANAGEMENT PHASE

Regardless of the species, the trees must receive sufficient weeding and tending, to remove weed and vine competition and to ensure that the trees get a good start in life. Where trees will be used for building materials, it is essential that vines are not allowed to climb the trees as they will ruin the shape of the stem. The amount of tending applied as well as the number of times, will depend on each individual's assessment of the situation.

Thinning will depend on what the individual wishes to use the trees for. The method of thinning may depend also on individual opinion, or on the prescribed technique for the species, and the amount of sunlight required for the crop planted beneath.

SEED ACQUISITION

To end this paper, a few pointers are given on obtaining seed. The seed of specific species can be obtained from overseas sources. In this case it is necessary to apply for an import permit, from the plant quarantine section of the Department of Primary Industry. The species name, the quantity of seed and the details of the seed source should be supplied with the request.

Seeds that can be collected locally should be taken from the best trees. That is they should be healthy and free of insect attack and diseases, and they should display the characters desired of the species. This may be fast growth, straight stem, light or heavy fruiting or seeding habits. Where possible good records should be kept on the seed source, the timber, collection and where it is being distributed, as either seed or as seedlings so that the species can be traced at a later date either for its source or origin, or in the field to assess its performance.

With research into plant reproduction being conducted by various organizations in the country, it is quite possible that seed of certain species of nitrogen fixing trees could be obtained from these organizations. Inquiries can be placed with the Director of Forests, with the Minerals and Energy Department, or with the Department of Primary Industry, if it is uncertain as to where seeds of specific species can be obtained.

Simple organic plant protection

Introduction

Here we are trying to present some simple methods of preventing or controlling pests and diseases in the garden without using chemicals. Some of these methods have been tried in the Demonstration Garden (WEI) and found to be very successful. Some of the examples given here are experiences from other countries but could be tried on our crops.

Many farmers believe that the only way to control pests and diseases in their garden is by using chemical pesticides, insecticides, fungicides etc. which they buy from chemical stores in towns. However, this is not always true and it is possible to produce good and high quality crops without using any of the chemicals.

Many of the chemical pesticides, insecticides or fungicides are not necessary and not suitable to be used by ordinary villagers or subsistence agriculture farmers or even workers of many farmers, because of the following reasons:

1. They are poisonous and dangerous to users, their families, their farm animals, wild animals, soils and waters.

2. They are expensive to buy and ordinary farmers cannot afford them.

3. They are difficult to get and not always available where and when you want them.

4. Our country's money is lost to foreign countries which sell these chemicals to us.

Our approach here in the plant protection is to help farmers create and maintain the balance of nature in their pieces of land. We want farmers to learn to use whatever is locally available and things they can grow themselves. We want to minimize the interference to their garden environment. Therefore, it is necessary to know about:

1. Correct plant feeding

2. Interplanting

3. Resistant varieties of plants

4. Garden sanitation

5. Cultivation

6. Timing of planting

The above are preventative measures or ways to avoid problems. Once problems begin control measures which can be employed are:

1. Biological control

2. Physical traps and scares

3. Biological repellents and poisons

Biological controls and cultural methods don't always get rid of pests and diseases completely, but can reduce them or their ability to damage the crops.

Healthy soil

It is claimed that "plants grown on rich organic soil, without chemical fertilizers, and without being sprayed, will not be attacked by insects because insects won't like their tastes". It is also claimed that "soils rich in organic matter tend to suppress diseases". "The diseases in soil will occur only when natural biological control is reduced by environmental disturbance, such as the application of certain chemicals in the soil. Plants grown on rich organic soil are always healthy and less attractive to pests and disease" (Roger, 1966).

You can help the beneficial organisms and encourage their work against bad organisms in the soil by:

1. Tillage — Tillage modifies the soil structure crumb size and helps aerate the soil. It reduces *Fusarium* (fungus) damage and facilitates root penetration.

2. Rotation — Crop rotation is the best method of controlling soil-borne pathogens (disease causing organisms) especially for cereals, potato, tomatoes and food crops. Crop rotation will reduce the build up of particular micro-organisms associated with a particular crop.

3. Trap crops and inhibitory plants — It is known that the marigolds inhibit root-knot nematodes. Asparagus inhibits stubby-root nematode from attacking tomato. Soybean residues control potato scabs, and barley residues control *Fusarium* root rot.

4. Adding organic amendments — Studies show that when manure and crop residues are incorporated into the soil, rapidly multiplying beneficial micro-organisms cause oxygen to be in temporary short supply. This weakens pathogens, which are overwhelmed by beneficial bacteria and digested. The better the soil is mixed with manure, the greater the percentage of digestion in a given time.

Control is best if you plant within a few days after addition of manure, then the pathogen spores have germinated and been digested and general microbial activity is still intense.

Planting at the time of adding plant residues results in increased disease. Planting too long after their addition also results in increased disease because the microbial activities have slowed down and pathogens are not attacked.

Cover crops and legume crops (particularly green legumes) when plowed under, have been an especially effective means of controlling various plant pathogens. Certain volatile substances released by decaying tissues may actually cause a higher germination of disease spores and at the same time prevent their growth and so they germinate and are quickly digested by soil micro-organisms.

Some examples:

1. Potato scab — ploughing under green rye
2. *Phymatotrichum* root rot — manure
3. Potato scab — grass clippings

Healthy plants

"Healthy Plants Resist Pests". Some healthy plants which sprout from organic soil are repellent to insects. It is known that insects thrive and reproduce better on unbalanced or inferior diets. While we are interested in proteins and minerals in our food, the insects go for plants having excessive carbohydrates. The following are a few points on plant nutrition and organic fertilizers:

Nitrogen:

— Aphids are said to respond positively to increased nitrogen. If you put too much fertilizer on your soil, you encourage aphids, including bean-, cabbage- and pea-aphids.

— Spider mites, on plants grown in low level nitrogen soils laid an average of 5.4 eggs/day, while those on plants on high nitrogen soils laid 10 eggs/day.

— Too much nitrogen weakens the plant tissue and makes the plant more susceptible to mechanical injury and diseases.

— In corn, N-deficiency shows in pale green to yellow colouring of initially the older leaves; it starts at the tips and progresses along the veins towards the leaf base.

— Tomatoes grown on high nitrogen soils drop blossoms without setting fruits.

— Balanced feeding Programmes:
A balanced natural fertilizing process will ensure against too much stem and foliage growth at the cost of fruits.

Phosphorus:

— "Master key to Agriculture" because low crop production is more often due to a lack of this element which can cause marked changes in pest behaviour.

— Spider mite egg production is encouraged by a phosphorus inbalance.

— Phosphorus deficiency can lead to problems with white flies in both outdoor and indoor potting mixtures.

— In corn, P-deficiency shows by producing of dark green, stunted plants with purple stripes.

To Correct for Phosphorus deficiency:

— Give the plants just the amount of phosphorus they can use.

— Add in the form of rock phosphorus, 30 – 50%.

— Bone meal, dried blood, cotton seed, soybean meal, activated sledge or basic slag.

Potassium deficiency:

— Yellowing, bronzing, reddening and death of the older leaves, starting at older leaves, progressing from tip along margin towards base.

— Mites thrive under such conditions.

— Too much potassium restricts the uptake of magnesium.

— Use manure or natural rock powders, the nutrients are made available to the plants slowly over a period of years.

OTHER IMPORTANT ELEMENTS

Sulphur deficiency:

— Chlorosis (Yellowing) of young leaves initially and later yellowing of all leaves.

Iron deficiency:

— Chlorosis of new leaves at the growing tip of the plants.

Magnesium deficiency:

— Chlorosis of young leaves mostly in alkaline or neutral pH soils.

Copper deficiency:

— Withering of leaf and stem tips.

Zinc deficiency:

— Caused by the use of certain chemicals (super phosphate).

— Yellowing between veins causes the dying of the tissue in tobacco, little leaf in citrus and leaf spot in sugar, bean and potato.

Boron deficiency:

— Dying of the growing tip of plant stems, internal cork in apple, water soaked areas and a bitter taste in cauliflower, and cracked leaf stalk in celery.

Natural fertilizers:

— Composted plant wastes.

— Manure.

— Nitrogen fixing trees, legumes.

— Free bacteria.

— Earthworms.

— Nematodes, millipedes, snails, centipedes (through their excrement and dead bodies). Pesticides and concentrated chemical fertilizers can make life hard and even impossible for these beneficial soil creatures.

Some organic fertilizers

Blood Meal:

— 15% N, 1.3% Phosphorus, 0.7% potash.

Dried Blood

— 12% N, 3.0% Phosphorus.

— Used direct or in compost pile.

— Use sparingly because of high N.

— Good for breaking down compost.

Bone Meal:

— 2 – 4% N, 22 – 25% Phosphorus.

— Melt or burn and crush.

— Effective with other organic materials.

— Controls potato bugs.

— Insect repellent.

— Repels aphid-carrying ants from garden, pots and flower beds by sprinkling a handful.

Compost:

— Best organic fertilizer.

— Contains all 16 elements or plant nutrients in the proper amount.

— Can include any organic material.

— Physical condition of the soil has an important bearing in the prevention and control of disease. Compost is the most important contribution you can make to your soil.

— A healthy dose of compost will keep tomatoes relatively free from blights and rots; fruit trees seem to have more resistance when well fertilized with compost.

Sowing seeds in a mixture of half compost and half sand will discourage 'damping-off' of young seedlings.

— The heat of a compost pile (70 – 75°C) can be sufficient to kill weeds, seeds and diseases on vegetable matter.

— Wire worms don't like composted soils as the organisms in the compost repel the pest.

— Compost has nutrients in the proper amount and gives the growers an insurance against diseases prompted by several nutrient deficiencies.

— Compost balances the soil by neutralizing it. Optimum pH is between 6–8.

— Counter an over acidic soil by adding limestone (not burnt lime), wood ashes, or shells to the compost pile.

Grass Clipping:

Because grass clippings are fairly rich in nitrogen they are used as 'green manure' to be worked into the soil, to be added to compost heaps and to be used as mulching.

Herbs:

— Any plants with repellent qualities, such as herbs, peppers, a number of the Allium (onion) family, can be used as mulch.

Leaf mold:

— To make nitrogen-rich (up to 5% N) fertilizer, spread the leaves and compost them in a container made of wood. Keep leaves damp and apply limestone.

— Leaf mold from decidious trees are richer in potash and phosphorus.

Leaves:

— Rich in humus and minerals including Ca, Mg, N, P, K.

— As compost, mulch, leaf mold.

— Controls pests and disease, e.g. tobacco, chillies.

Limestone

Several common diseases — including potato scab, beet rot, tobacco, tomato wilt thrive in acid soil and are consequently reduced by the neutralizing action of limestone. The micro-organism that produces penicilin, streptomycin and other soil antibiotics, which in turn kill or make harmless the microbes causing these diseases, must have calcium and magnesium to do their work. Acid soil supplies these nutrients, and therefore increases the production of antibiotic-producing microbes. Besides increasing the fertility of the soil, crushed limestone can be built up around tree trunks to act as a shield against mice, snails and slugs.

Dried fresh manure

Fresh manure is too rich for some applications, because of the high nitrogen content. It is best to allow manure to decompose into compost before applying directly to the plant.

Mulches

Mulch has been recognized as highly beneficial to plants because it helps in maintaining uniform soil temperature.

The root systems will stay cool and moist during dry season and are kept warm in cold areas.

Mulches are also said to discourage diseases and pests by producing antibiotic agents that are picked up by a plant's root system. Heavy mulching reduces root knot injury caused by nematodes, black spot in roses (a disease spread by splashing of water) and tomato blight (a viral disease).

Grasshoppers are stopped by a layer of mulch — grasshoppers have a hard time hatching from their eggs. Troublesome weeds such as morning glory, thistle and quack grass can be controlled by covering the weeds with a smoother mulch made of organic matter and waste paper.

1. *Newspaper and magazine papers*

Can serve as soil builders and animal repellents. Repeated use on the garden will tend to make the soil acid, and so lime should be applied at suitable intervals. Add nitrogen in the form of blood meal to assist decomposition of newsprint. Blood meal serves as an animal repellent in its own right. A half (½) inch layer of newspaper is said to prevent root knots on okra crops. Attention: The ink of the newspapers is poisonous.

2. *Peat moss*

Does not contain nutrients but serves to aerate the soil, improve drainage, and help plants to absorb nutrients from the soil.

3. *Sawdust*

Use well rotted sawdust only, as raw, pale-coloured sawdust will mat and cake, thereby preventing proper penetration of rain water. A half (½) inch of sawdust mulch will keep down the weeds in a vegetable garden and also conserve soil moisture. A 5–6 inch band over the rows of seedlings can prevent soil crusting but cultivation is required between rows to get rid of weeds. Supply plants with organic matter such as cottonseed, soybean or dried blood or bone meal to prevent nitrogen shortage which sometimes develops using a sawdust mulch compost.

4. *Wood ashes*

This is good alkaline fertilizer containing 1.5% phosphorus and 7% or more potash. If left unprotected in the rain, wood ashes will lose much of this potash. Wood ashes can be mixed with other fertilizers, side dressed around growing plants or used as mulch.

By spreading a circle around the base of the plants and the soaking area you can control maggots, cutworms, cucumber beetles, squash borers, red spiders, potato bugs and slugs. Don't allow the ashes to contact germinating seeds and sensitive stems and roots.

5. *Wood chip*

In addition to aerating the soil and increasing soil moisture holding ability, woodchips have a higher nutrient content than sawdust.

6. Some other crop and animal wastes which can be used as mulch and to fertilize the soil:

a. Hoof and horn meal
b. Dry fish scraps
c. Feathers
d. Brewery cannery wastes
e. Hay and straw
f. Tobacco stems
g. Lobster shells
h. Molasses residues
i. Garbage

Soil moisture

With many plant-insect relationships, either an excess or a deficiency of soil moisture may make the plant more susceptible to insects. Organic matter (humus) in the soil helps protect trees against drought and acts as a reservoir to balance soil moisture reserves.

Ants do not like wet conditions in the soil and ants are the agents behind much aphid damage. Many physiological problems will clear up as well.

Blossom rot of tomato is brought on by insufficient soil moisture. One recommended remedy is to mulch the plant. If adequate moisture is not available in the soil, the plants will remove water from the tips of the fruit, bringing on the end-rot.

Protecting crops with other plants

There are four types of cultural controls possible, to prevent pests and diseases, by growing certain plants next to each other:

1. Companion planting

2. Mixed cropping

3. Repellent planting

4. Trap cropping

Companion planting

Companion planting is when crops are in "happy combination". This has been learned through observation. Combinations of plants produce crops that not only grow better, but are relatively free from insects and diseases. The protection may be due to the increased health of vegetables growing near their 'friends' (Table 3). As stated earlier, insects typically pass by healthy plants, preferring to feed on weak or sick plants. Why is companion planting healthier? Perhaps because the plants make complementary demands on the environment.

Table 3: Example of useful combination of crops

1. Celery and leaks

2. Marigold and tomato or other crop plants susceptible to nematodes (Fig. 139)

3. Mustard and plants susceptible to low pH

4. Mustard and plants susceptible to nematodes

Fig. 139 Combination of tomatoes and marigold plants. Marigold is protecting the tomatoes of being attacked by nematodes.

5. Some deep rooted plants, e.g. *Leucaena* had the ability to bring many essential minerals towards the surface making them available to the companion plant.

This is important in plant protection because plants with deficiencies are especially prone to diseases and insect attack.

Methods of planting a Companion Garden

When planting a companion garden, it is good to get the companion plants close together:

1. Zigzag rows eg. zigzags of onion and beets tucked into one another

2. Intercropping — plant several companions in the row

3. It is best to place paths between sections rather than between companions

4. Plant herbs on borders and scatter them throughout the garden.

Mixed cropping

Mixed cropping is when many different crops are grown in a mixed fashion on a piece of land. In gardens, as in nature, plants have better change of survival if they are surrounded by plants of various species. Plants in monoculture, or only one type of crop grown in one big area, are more susceptible to their enemies. A good mix of plants encourages a healthful variety of insect life. The principal mechanism at work here is the perpetuation of natural controls, such as predators which kill many plant pests. Beneficial insects typically need alternate hosts and if you have several different plants growing nearby, you'll have beneficial insects on hand in case of pest attack.

Repellent planting

Repellent plants bestow some protection on nearby crops, up to a distance of one metre. Some repel a wide variety of insects and others are effective against a particular pest. Garlic, mint, and marigold are known for their built-in repellent to insects. A number of other plants have this trait too.

Table 4. A list of pest repellent plants.

PESTS	REPELLENT PLANTS
Ants (and aphids they carry)	— Spearmint, penny royal, tansy, southern-wood
Aphids	— The above as well as: garlic, chives and other alliums; coriander, anise, nasturtium and petunia around fruit trees
Asparagus beetle	— Tomato
Borer	— Garlic, tansy and onion
Cabbage maggot	— Mint, tomato, rosemary, sago, planted in adjacent rows
Cabbage moth	— Mint, hyssop, rosemary, southern-wood, thyme, sage, wormwood, celery, catnip, nasturtium
Carrot fly	— Rosemary, sage, wormwood, black salsify, various alliums
Chinch bug	— Soybeans
Colorado potato beetle	— Green beans, horse radish, dead nettle flax
Cucumber beetles (spotted and striped)	— Tansy, radish
Cut worm	— Tansy
Eelworm	— Marigold
Fruit tree moth	— Southernwood
Leaf hopper	— Pefunia, geranium
Mexican bean beetle	— Marigold, potato, rosemary, summer savory, petunia
Mice	— Mint
Mites	— onion, garlic, chives
Mole	— Spurge, castor beans, mole plants, squill
Nematode	— Marigold, salvis, dahlia
Slugs (snails)	— Prostrate rosemary, wormwood
Squash bug	— Tansy, nasturtium
Striped pumpkin beetle	— Nasturtium
Tomato hornworm	— Borage, marigold, opal basil
White fly	— Nasturtium, marigold, nicandra (previously called perruvian ground cherry)
Wireworm	— White mustard, buckwheat, wood

Trap cropping

Some companion plants are highly attractive to pests and lure them away from valued crops. These are called trap crops. In the garden at Wau Ecology Institute the following observations were made:

1. Dill detracts tomato harmworm away from tomato.
2. White or pastel zinnia attract Japanese beetles from valued crops.
3. White rose attracts Japanese beetles.
4. Odourless marigold attracts Japanese beetles.
5. Knot weed attracts beetles if planted at the edge of the garden.
6. Soybean attracts Japanese beetles.
7. Mustard planted as early as catch crops saves cabbage from harlequin bugs.

Once pests flock to the trap crops you can collect them and kill them. Thus these plants can serve as little insecteries. With plenty of food at hand, your valued crops will be off to a good start.

Use of resistant varieties

Resistant varieties are varieties or strains of plants that are observed to have fared much better against insects or diseases than an adjacent variety of the same crop. These resistant plants can be categorized into:

1. NON-PREFERENCE — GROWING UNAPPETIZING PLANTS

A number of things that may turn insects away are:

a. *Colours:*

Red cabbage shows resistance to loopers and imported cabbage worm. Yellow-green pea varieties are more resistant to pea-aphids than are blue-green peas. Insects are colour blind to red and yellow coloured plants.

b. *Hairs*

Hair in plants offer some resistance.

c. *Odour*

The aromatic oils or esters in some plants repel insects or resist insect attack.

2. ANTIBIOSIS

These are plants that have the ability to prevent, injure or destroy insect life. The insects taking these plants might vomit or go into a stupor, become stunted, have poor fertility or die. Antibiotic plants contain some valuable oils that repel insects or fungus. Some plants seem to produce those oils when under attack. Some plants such as garlic and marigold have a lot of these bitter tasting oils.

3. RESISTANCE AND TOLERANCE

Some plants can endure pest attack. In contrast, to the other two types of resistance, a tolerant plant survives and produces a healthy crop. You can plant varieties that are also resistant to disease. The disease resistant varieties are easier to come up with, than those resistant to insects because diseases generally tend to stick to one species more than insects. Insects, if they are repelled by one plant usually go to another plant.

So, it seems unlikely that you can grow a crop totally free of pests and diseases. But resistant varieties are one more way in which to grow your food without using poisonous chemicals.

Taking advantage of insects' natural cycles

Timing, tillage and mowing to interrupt insect cycles

It helps to keep a calendar of gardens events: when pests first appear, when they are at their highest numbers and when they subside.

By knowing the pest's life cycle you can anticipate trouble by tilling to catch them when they are at soil-borne stage or growing crops when the pests are not there.

Natural occuring predators and parasites

Most of all plant-eating insects (potential pests) are kept in check by their natural enemies, predators and parasites. Dust and spray will kill beneficial insects and animals that feed on pests. Chemical pesticides also make pests become more resistant to these chemicals and so later the chemicals are useless and have no affect on the pests.

So it is best to let nature maintain itself as it has for millions of years.

Table 5. List of some common predators and parasites.

PREDATORS & PARASITES	PREYS
Ground beetle (caterpillar beetle)	— Gypsy moth, larval stalk canker-worm, army worms and cutworm
Lady bugs	— Larvae feed aphids
Sting bugs	— Cankerworm, cutworms, colorado potato worm
Blister beetles	— Egg pots of grasshopper
Firefly beetle	— Larvae feed on slugs and snails
Ambush bugs	— Any insect including beneficial wasps and bees
Assassin bug	— Caterpillars, leaf hoppers, Japanese beetles etc.
Damsel bugs	— Aphids, leafhoppers, treehoppers and small caterpillars
Flies	— Parasitizes and preys on various insects
Tachnids	— Corn borers, larvae of brown tail and gypsy moths
Flower and robber flies	— Adults pollinate crop flowers, larvae feed on aphids, leafhoppers and mealybugs

Nematodes	— Carnivorous nematodes prey on other nematodes and other pests
Lace wings	— Aphids and mealy bugs, larvae live on thripes, mites, caterpillar eggs, scales, leafhopper, nymp, aphids and mealy bugs
Ant lion	— Larvae eat ants that fall into its pit
Damsel flies & dragonflies	— Mosquitoes
Praying mantis	— Feeds on any insects
Spiders	— Many insects
Wasps	— Parasitize the larvae of most moths, sugar cane borers, cutworms and larvae of many beetles, mealy bugs, aphids and scale insects
Bacteria	— Various insect diseases, insect larvae
Virus	— Obligate parasites of larvae of various insects. Mass produced for insect sprays
Fungi	— Responsible for a number of insect diseases
Birds	— Many birds eat a number of potential insect pests
Animals	— A number of insects eg. Lizards, frogs, eat insects

Traps and barriers

Insect and pest traps

Pests can also be reduced by luring them into traps. Most traps work by appealing to an insect's need for food, shelter or a mate. Simple traps that you can make yourself are better than commercial types, because simple traps are less apt to cause a mass killing of good insects along with the bad. Some of the pest traps are as follows:

1. *Snail and Slug Trap*

    a) Using a bit of stale beer in a shallow plate or container, the slug or snail will crawl into the liquid and drown in it. If they just drink and then leave, mix a bit of flour with the beer to make a sticky mixture.

b) An inverted cabbage leaf placed on the ground will attrack snails, slugs, cutworms and other pests that hide during the day and forage at night.

2. *Earwig Trap*

a) Fill hot pork grease into a 4–6 ounce can until ¼ full, place outside where needed.

b) Invert a flower pot filled with moss on a stake. The trap should be examined every morning and evening and the earwigs should be destroyed.

c) Tie 4 pieces of bamboo, 30 cm in length with both ends opened, in bundles. Paint light green and place in places needed. The attracted earwigs should be shaken into a bucket of water with a layer of kerosine.

Earwigs and various beetles will hide under these boards laying in the garden.

3. *Fruit flies*

Citronella oil attracts only male fruit flies.

4. *Ants*

Can be lured into containers baited with sugar water, fats or any other food residue.

5. *Grasshoppers*

Attracted by all kinds of scents eg. malles, citrus fruit, lemon or vanilla extracts, beer, vinegar, salt, soap and smoke.

6. *Traps in Trees*

1. *Codling moth*

Mix molasses, water and sugar in a bucket and hang on trees.

2. *Light Traps*

Flying insects can be attracted by light. Some colours of light are more attractive than others. For example; Black light (ultra violet) attracts adult moths of: Tomato and tobacco hornworm, codling moth, oriental fruit moth, corn earworm, corn borer, cabbage looper, fall armyworm, bagworm, spotted cucumber beetle, cutworm and a hundred more.

Red, orange and yellow lights are avoided or ignored by almost all insects. A regular light bulb or lamb attracts some pests including the corn borer. An easy way to make light insect trap is to hang a lantern on 3 poles with the top ends, tied together as shown in Figure 140.

Fig. 140 Simple kerosine lamp as light trap for insects

Fill the bowl with water and a layer of kerosine. The insects attracted by light will try to land on the lamp and fall into the container of water. Kerosine helps to kill these insects.

Protective borders and barriers

1. *Protective borders*

Instead of applying repellents and insecticides right on the crops, you can set up lines or rings around patches of certain plantings or even the entire garden. Wing insects will fly over such barriers but those which crawl or walk to eat these will be stopped.

a) *Wood Ashes*

A perimeter of wood ash scattered around the base of plants will block all sorts of walkers or crawlers. You can also sprinkle lines of lime, bone meal, powdered charcoal, or any vegetable dust.

b) *Boards*

Set boards about 10 cm high around the lettuce bed and paint with fuel or dump oil. This stops crawlers from getting to the crops.

c) *Papers, Plastic Bags and Leaves*

Fruits like bananas, guavas, etc. can be protected from insects, birds and other animals by wrapping or covering the individual fruit or cluster of fruits in paper bags, plastic bags or leaves. Covering may also help the fruit colour and help them ripen quickly and may even increase the size of individual fruits. If you have money you can buy nylon nets, or wire nets to cover the entire tree to protect it from birds and other animals.

2. *Barriers*

Barriers are a type of trap that do not attract insects but insects that just happen to wander into it are caught. Some of the barriers are as follows:

a) Deep furrows between rows around patches

Crawling insects such as chinch bugs, army worm etc., are halted in the furrows and killed by pouring a little bit of kerosine, lime or they can simply be buried.

b) Sticky bands can be applied to tree trunks and stems of large plants to halt the advance of a number of crawlers. There are commercial preparations tapes such as "Tangle foot and stiken" but you can prepare your own by mixing pine tar and molasses and oil. Such sticky preparations may damage the trunks, so it's a good idea to lay down a layer of cotton batting or heavy paper before applying them.

Repellents and poisons

Natural sprays and dust: The last type of defence.

Previously we have suggested ways of saving your crops from damage. This one should be used at the point when you can no longer tolerate the damages and diseases that are taking place in your garden.

The options are of two types: 1. repellents, that cause the insects to avoid the treated plants; and 2. insecticides, that kill the insects. It is best to chase away the insects by repellents instead of killing them with insecticides because all insecticides kill both bad and beneficial insects. It may be a good idea to identify the enemy properly before deciding what spray or dust to use. It may turn out afterwards that the enemy causing the problem might be controlled more easily by other biological control measures than by the sprays. The hot or bitter tasting, or strong smelling plants such as garlic, peppers (chillies) and many other herbs have been discovered to be powerful insect repellents. These plants can be grown in your own garden and you can make sprays yourself. There are powerful organic sprays and powders available commercially: For example, ryania, votenone (deris), and pyrethrum — but those you make at home are tamer and cost nothing.

Insecticides made from garlic are non-persistent substitutes for DDT and other chemical insecticides. A garlic insecticide is not toxic, even in very high concentrations. As well as being an insecticide, it also has antibiotic powers which can be used to fight various plant diseases such as downy mildew of cucumber, bean rusts, early blight of tomato, bacterial blight of beans. Other alliums are nearly as powerful as garlic.

Recipes of sprays and remedies

Garlic

Ingredients:

3 ozs chopped garlic

2 teaspoon of mineral oil

1 pint of water

1 oz of pure soap (not detergent)

Method:

Soak chopped garlic in mineral oil for 24 hours.

Slowly add soapy water and stir well.

Strain liquid through fine gauze and stir in a china or glass container.

For all pests and diseases:

Dilute 1 part to 20 parts of water or 1 part to 100 parts of water.

Variation to this:

½ cake octagon soap

2 litres hot water

2 mashed garlic bulbs

4 teaspoons red pepper

The spray will cause tree caterpillars to fall to the ground.

Onion

Ingredients:

1. 6 onions mashed and simmered in a large container for several hours makes a strong soup and spray.

   This is a spray for diseases, especially of fruit trees.

2. The simplest onion remedy for vegetables disturbed by cutworm is to tie onion leaves on the stems of vulnerable plants.

Pepper

A typical recipe is the combination of pepper, garlic and onion in the following manner.

1 clove garlic	)	Blend in water as a soup and allowed
2–3 red peppers	)	to sit for two days. Then strain and
½ wild green pepper	)	spray.
½ onion	)	

Spray all plants against ants, spiders, cabbage worms, caterpillars and tomato hornworms. Pepper spray can also control a number of viral diseases such as cucumber mosaic, ringspot and tobacco.

To chase ants from the ground near the plants sprinkle red hot pepper onto their hills.

Recipes from herbs and flower tea sprays

Many of the companion plants that protect nearby plants can be dried and a tea made from these plants make excellent sprays.

1. 4 parts desert sage
   2 parts erigeron
   2 parts eucalyptus
   2 parts silky wormwood
   1 part yerba santa
   1 part wild buckwheat

2. Lemon grass + hot pepper

3. Chamomile tea — prevents damping off on seedbeds and flower pots
— prevents mildew of cucumber seedlings

4. Tomato leaf tea for green flies and caterpillars

5. Wormwood tea for slugs, aphids, other tender-bodied pests

6. Stinging nettle tea for aphids.

Recipes from the kitchen and workshop

1. Soapy water and soap suds

   Naphtha soap is recommended. Soap will dissolve well if you add a teaspoon of alcohol and a quarter of water base. This is for mealy bugs, leafminers, lace bugs, aphids, leafhoppers and mites

2. Soap and kerosine

3. Fish oil soap against larvae

4. Milk and water (9 parts) — Tomato mosaic, tobacco mosaic, sugarcane mosaic

5. Powdered sugar — worm

6. Powdered sugar and wheat flour — cabbage worms and moths

7. Salt — a shake or two will kill a slug

8. Coffee — a cup or two of strong coffee will control red spider mites

9. Asparagus — The juice will control nematodes

10. Tobacco — spray and soap will kill most insects. Be sure to rinse off the spray after treatment so that the foliage is not burnt.

Women in agriculture

Introduction

The part that women play in subsistence agriculture is of vital importance and contributes directly to the survival of the family. Feeding the population, in all countries, is an important national function. It is also true of Papua New Guinea. However, both national and provincial governments, do very little to improve the subsistence agriculture which is the nation's foundation for survival. Women in Papua New Guinea contribute significantly in this vital, but nationally neglected subsistence economy.

The large majority of our population derive their livelihood from subsistence agriculture. Therefore, Papua New Guinea should come up with plans and programmes to introduce simple technology that will increase production capacity and out-put in subsistence farming. It should be aimed at improving nutritional standards, better and improved farming methods and techniques and to uplift the quality of life of the bulk of the population.

The role of the women in the traditional society

Papua New Guinea is a diverse society in which there are over 700 different languages spoken. There are many different tribes, with different traditions, beliefs and values. However, there are certain common norms which define the expected roles of males and females.

Traditionally there is a division of sex roles, where men and women are expected to perform certain tasks. Generally, women's primary roles are to produce food, rear children, care for domesticated animals etc., while men are expected to perform the more muscular tasks such as clearing up new garden sites, building houses, maintaining fences, hulling out canoes, fishing or hunting, etc. In a traditional society, everyone from a child up to adulthood has particular duties to perform for the well being of the whole community.

Women were not traditionally treated or even thought of as being an inferior member of the society. Women in Papua New Guinea played a very important part in the community. They produce food to feed the family and they bring up the children. Women are the "bread winners" of the family, unlike the western tradition where men are the "bread winners".

Subsistence agriculture patterns

Throughout Papua New Guinea, subsistence agriculture practices are basically similar. In the past, every member of the family had clear roles to play. When a new garden was made, children, youths and adults worked together, each knowing what they had to do.

Men and youth would slash and burn a new garden site. They would then collect all the big pieces of wood and pile them either on the side of the garden or in the cleared area, to be used as firewood when they were dry. Women and young girls would collect leaves and smaller branches to pile them up and burn. Men would dig out smaller tree stumps to produce more planting space, leaving behind huge stumps. Neighbours and relatives also helped. In return when food was harvested, the owner of the food shared their harvest with neighbours as "payment" for their help. Relatives normally would be given a portion of the garden to plant their own food crops.

In the Highlands it is a normal practice to dig drains in the gardens. In water logged areas, drains were dug to get rid of excess water and to serve as drainage.

Normally, clearing and burning of garden sites are done during the dry season. While clearing is done, seeds and planting material are prepared. Planting begins after the clearing of garden sites. There are some food crops which men alone are responsible for planting. Likewise, other crops are left for the women to handle. Girls work closely with their mothers and elderly women while young boys work alongside older men. For the youth its a learning process, an educational experience where ideas, methods, and gardening rituals and techniques are passed onto the younger generation.

In the Highlands men are responsible for planting sugar-cane, bananas, pitpit while women are responsible for planting corn, green leafy vegetables, beans, cucumber, pumpkin etc.

Weeding and general tending of the garden is done by women and girls. Harvesting is also a function performed by women as is the food preparation for consumption. Bananas and sugar-cane take a longer time and men tend to care of these crops.

For a sweet potato garden, farmers have their own criteria for site selection. Normally men clear and make drains or they make moulds. In higher altitude areas moulds are jointly done by men and women. Weeding, harvesting and cooking is primarily women's responsibility.

In a traditional subsistence agricultural society, every member of the family is occupied in doing something. No one is a drop-out in the village system; everyone feels a part and participates fully in the social and economic activities of the community.

Apart from producing food for the family, women also are responsible for children's welfare and up-bringing. Women pass on to children the morals and whatever the children needed to know at an early age. Both males and females grow up together in their early childhood period.

When the boys reach ten years and up to puberty, they accompany their uncles and fathers, while girls of this age accompany their mothers in communial or village chores.

Western influence on women and subsistence agriculture

During the colonial period, there was much emphasis on cash cropping and large scale livestock development projects. These agricultural patterns and systems were very new to individuals who had had a long history of subsistence gardening where slash and burn was their traditional system.

Introduction of western christianity into the communities changed life style patterns, beliefs, attitudes and behaviours. Tribal warfares, and ceremonial activities have been stopped, due to western christian influences. People believed that the christian belief or western christianity was a better way of life than their own. Western missionaries went throughout the country converting people to believe and practice christianity in the way the missionaries perceived. Followed by the colonial government administrators, who came in with their own system of government, missionaries and government officials played a complementary role in the early days to take over Papua New Guinea. Colonial expansion, exerted a significant impact upon subsistence agricultural life.

In some communities, men were taken out as labourers to work on cacao, copra and rubber plantations, far away from their own village environment. This resulted in women assuming more work, doing what they used to do and at the same time performing the work normally done by the men. Women had no choice but to adjust to changing circumstances in order to survive. Women had to produce enough food to feed their children under the subsistence level of food production.

In Papua New Guinea's urban cities and towns, there exists a sector of women who have migrated either accompanying their husbands or younger village women who leave the village to seek 'a better life'. Women whose role has been a subsistence food producer are now exposed to a very different environment. They either live in settlement areas where it is overcrowded with many different people from all over the country or in a housing commission area with very limited space for any gardening activity. She is put in a situation where she is dependant and feels isolated.

Some village women in urban areas have occupied themselves in gardening on vacant government land. In Lae, the adjoining Atzera Hills have been over cultivated, causing landslips, excess water run-off and destruction of the environment.

The quality of food produced in these areas is declining due to land deterioration, loss of soil fertility and infestation of destructive insects. Women are continuing to make gardens to keep themselves occupied and to supplement their husbands low income and sell some of the produce in the local market. These activities are not the same as what she would do in the village but at least she is trying to exist in an urban environment.

It has been noticed that the best food is sold in the market and only the rest is consumed by the family. Therefore, the food quality is low and the meals are normally unbalanced, resulting in malnutrition problems for the children and the family. Women from the rural areas think that everything in towns and cities is good. Therefore they very often spend their little money on junk food like cheesepops, biscuits and loliwater. People in general have the false notion of western civilization as being superior to their Melanesian tradition.

Agriculture extension training was geared to producing manpower for cash cropping and large scale livestock production. It follows the path layed by the colonial administration. Before independence, only men were recruited in Agriculture Colleges. The colonial administrators had by-passed women, because in their own countries, women were not taken as co-partners, with responsibilities just as important as men. Thus the treatment of women as inferior in this country can be attributed, to a certain extent, as the result of colonial policy.

Many western trained academics, anthropologists etc., have interpreted women in Papua New Guinea to be beasts of burden, which is not true. Women's role within the community in Papua New Guinea has been totally misinterpreted by these out-siders.

During the colonial rule, the Department of Agriculture Stock and Fisheries, introduced new cash crops such as coffee, cacao, copra, rubber, etc. People were told that they could get money for the crops. Very often villagers established cash crops on their garden lands near their homes. This meant that the women had to travel some distance to cultivated land to produce their family's food. Large scale livestock production, especially cattle, under village conditions had a lot of problems because expatriate cattle projects, on either government land or land purchased from village people, had modern technology to assist in continuous and better quality production than what a villager could produce. Villagers had to divide their days to do community activities and food production and at the same time manage whatever projects they established. Women also spent some time working on the cash crops, besides their normal duties.

In 1975 when Papua New Guinea became independent, female students were recruited into Agriculture Colleges to undertake agriculture courses and graduate as extension officers. This was done, hoping that there would be more contact with women who were involved in cultivating the land. These female extension officers seemed to have fallen into the same trend as their male counter-parts, just travelling to see and advise in cash crops. However, there are a few female officers who are at least attempting to work with women at the subsistence level. It seems that individual attitudes, commitment and the ability to adapt and accept village women as they are is the key to extension work for female extension officers.

Women in Papua New Guinea are experts in gardening in their own way. For new practices to be accepted one has to prove to them, in physical terms, the benefits of a good harvest. This means that one has to wait for a whole cycle to prove that the advice given is valuable. Many a time, extension officers introduce particular crops and they do not return. It has been shown that the extension system is ineffective in demonstrating techniques that are useful.

Subsistence agriculture has never been a priority in government development plans. Politicians have been talking about improving subsistence and village conditions. However, in terms of work and government finance, there has not been anything done in improving and introducing simple and appropriate technologies that will increase subsistence food production, until recently in only one province of Papua New Guinea. The majority of the Papua New Guinea population still live on food production by subsistence gardening.

Problems women are facing in subsistence food production

Today in many parts of Papua New Guinea women are having problems in producing good quality and enough food to feed their families. Population pressures have forced people to over cultivate their land, which has led to loss of soil fertility and erosion.

In other areas where population is not so dense, women are walking far distances to cultivate good fertile soil to plant their food crops. Land close to the villages are no longer suitable or is cropped with cash crop or it has developed into grassland due to regular burning. Women are facing social problems such as men spending their time consuming alcohol, and neglecting their tasks and the gardens. Fatherless children are being born in many villages today, because the traditional morals and strict behaviour are not followed any more. Changes have become more diverse and rapid in Papua New Guinea and they directly affect traditional cultural structures and norms.

Therefore, the big changes are not desireable in the improvement of subsistence farming but there is a demand for the innovation of simple appropriate techniques which subsistence farmers can integrate into their present farming methods. These techniques must improve food production, prevent further deterioration of the land resources, help the village farmers to survive, restore dignity and help them to adapt to destructive changes caused by big development projects sponsored by government and foreign investment.

The position of women in Papua New Guinea today

Although the position of women is slowly improving, they are separated within Papua New Guinea into four distinct levels, according to their education:

1. Women in the villages who have not attained any form of formal education either by missions or government organizations.

2. Women who are semi-educated living either in villages or in urban areas.

3. Women who have attended formal education and have attained certificates or diplomas and are employed either by the government or by the private sector.

4. A few women that fall into the elite group, which consists of university graduates and women holding top positions in the government or the private sector.

The subsistence agriculture is dealing especially with women on the village level which have not received any formal education. It is their problems that have to be solved, including soil fertility problems, changes in living standards, timing of the various duties like collection of firewood, collection of water, household requirements and other non-agricultural activities.

Different government officers often come to villages to deal with the same people and the same subjects. They take up too much of the time of the village people. This could be made more effective by arranging the place and time on which all officers on one district will be available to the villagers for help.

Furthermore, it is essential for the effectiveness of the extension work that the villager knows to whom he has to turn in case of arising problems. The Morobe Womens Association for example is working to be such a group to whom the villager can refer to instead of having to sort out a dozen different specialists. Wherever possible, already existing groups, like church organizations should join the association, thus providing the core with their extension knowledge.

Whatever is done, it is impossible to go back to the traditional system as it was. An appropriate system that gives every person access to the basic needs like food, housing and clothing must be looked for.

Application of work study techniques, points about extension methods and nutrition improvement in subsistence farming

To give some ideas how the problems involved in the improvement of subsistence agriculture by extension officers can be approached a short and by no means comprehensive compilation of some techniques and methods are given. These may be helpful to extension officers for their duty in the villages.

The application of work study techniques to improve subsistence agriculture systems

INTRODUCTION

As for other systems, the tools and methodology used in work study systems can be used to analyse the problems involved in improving subsistence agriculture. Only two systems, the ideal systems concept and the critical examination sheet, will be discussed here.

THE IDEAL SYSTEMS CONCEPTS

In all production systems, whether it be in industry in the developed nations, or in subsistence farming in the developing countries, the following should be observed:

1. There is a present cost to every unit of production
2. The ideal cost would be zero (theoretically)
3. Designs for the ultimate ideal system could be worked out in the designs room
4. The technologically feasible ideal system has costs in between the present costs and the calculated costs of the ultimate ideal system.

The diagram of the ideal system concept serves as a constant reminder that with time, every system could be improved with a change in the circumstances which always take place in our world (Fig. 141).

This improvement is reflected in reduced costs and improved quality. To those of us who have not been exposed to the ideal system concept, the achievement of zero cost may seem to be an impossible task. But nothing is impossible to human ingenuity. There is even a report of negative cost beyond zero.

Take for example the water-can rose fiasco. The rose used for an even distribution of water over small seedlings may be efficient for developed nations who have pipe borne water without any debris in it. But it is a catastrophe, when used with stream or well-water which has debris floating on it. The debris in the water the subsistence farmer in a developing nation uses, blocks the holes in the water-can rose. He dislodged them very ingeniously by knocking the rose on a hard surface. But there is a limit to the beating a rose can take. It is eventually beaten out of commission.

This water-can rose has now been replaced by us with an imported piece of tin; the top of a circular meat can (lid) which

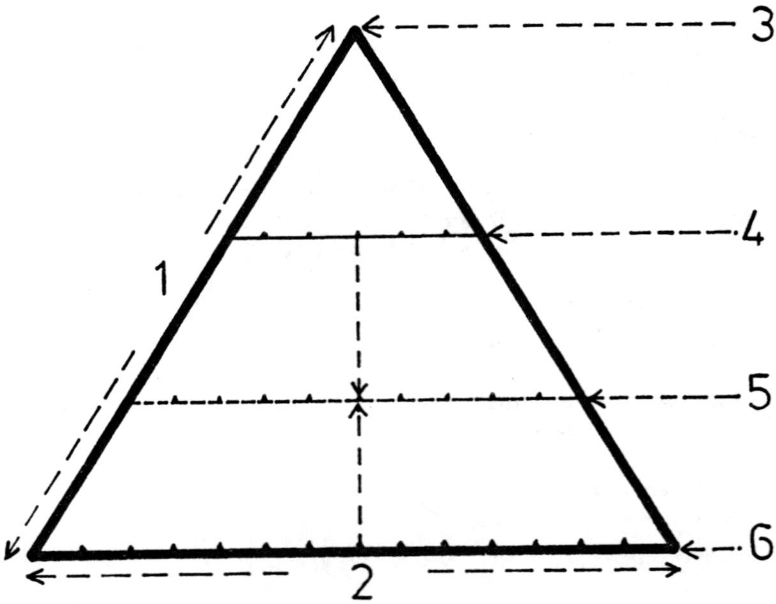

Fig. 141 Diagram of the ideal system concept
1 Time, 2 Costs, 3 Theoretical ideal system, 4 Ultimate ideal system, 5 Technologically feasible ideal system, 6 Present system cost

comes along with the garbage we use for composting. It works better than the costly rose. More, it has a built in control that impels the worker to greater speed than before. Also, it has brought back into commission water-cans which had been retired for lack of the rose.

The procedure on how to make this type of lid-rose is described under 4.6 of this manual.

We find that this device, a baffle we should say, eliminates all the draw backs found in the water-can rose. It can be made by any unskilled worker or villager and it costs just about 2 minutes of work-zero costs at the best.

THE CRITICAL EXAMINATION SHEET

Ruyard Kipling was really wise and far-sighted when he wrote this:

I have six honest working men,
they taught me all I know,
their names are What and Where and When,
and How and Why and Who.

The critical examination sheet uses these six working men in a ruthlessly, analytical way. Nothing is left to chance. Every detail is analysed (Table 6).

Table 6. Method study critical examination sheet

Date of study		No.	
EXISTING/PROPOSED METHOD	WHY	ALL ALTERNATIVES	CONSIDER FOR DEVELOPMENT
FUNCTION What is achieved? (1)	Is it necessary? (2) Why? (3)	What else could be done? (12)	Eliminate if (2) is NO. What should be done? (17)
EQUIPMENT How is it done? (4)	Why that way? (8)	How else could it be done? (13)	How should it be done? (18)
SEQUENCE When is it done? (5)	Why then? (9)	When else could it be done? (14)	When should it be done? (19)
LOCATION Where is it done? (6)	Why there? (10)	Where else could it be done? (15)	Where should it be done? (20)
PERSON Who does it? (7)	Why those persons? (11)	Who else could do it? (16)	Who should do it? (21)

Note: This form should be expanded to about 2 feet to allow plenty of entries by the brainstorming group. The sequence of the group's agenda is that of the numbers shown for each cell.

Some useful points about extension methods

Educational programs which reach beyond the campus of a college, school or university are called extension programs.

A special branch of extension is agricultural extension. This means an educational program in rural, suburban or urban communities relating to farming, animal husbandry and related agro-based industries.

It is usually performed by so called change-agents of governments or voluntary agencies. Usually they are called extensionists or extension officers. A change-agent is any officer of a government, or non-government agency who works on education programs in communities, with a view to bring about improvements in education, health, agriculture and other economic activities which will raise standards of living.

Skills required in a change-agent are:

— Resource management competence
— Man management skills
— Public speaking and salesmanship
— High knowledge of his own work
— Reporting and record keeping abilities

Attributes required in a change-agent are:

— Good attitudes towards people and work
— Behaviour which will not be questionable
— Commitment to organization goals
— Dedication to work up to a point of self-sacrifice
— Energy sufficient to keep him going more than 8 working hours a day
— Punctuality in attending to appointments
— Substained interest in projects undertaken
— Continuous expectations and achievements of the best result.

Methods which should be or can be employed to reach the optimal goal in extension work are:

— Employment and training of village extensionists
— Regular training of selected village groups
— Result and method demonstration plots
— Trial plots
— Distribution of proven seed materials
— Assisting in grading and marketing of produce
— Organization of field days and exhibitions
— Radio programs and newspaper publicity to make people proud of their doing
— Printed handouts
— School and home gardens.

Points for improving the nutrition of rural people in Papua New Guinea

According to research findings, the protein intake of people whose staple food is protein-poor kaukau, can be improved in two ways:

1. By eating more good varieties of sweet potato. Children could be made to eat more through the use of locally prepared pickles, spices and chutneys added to the sweet potato.

2. The use of unpeeled sweet potato.

Therefore, the following points should be held in mind during extension work in addition to recipes and specific cooking and eating habits described extensively elsewhere:

— Cultivation of protein rich sweet potato varieties and protein rich maize and bean varieties
— Cultivation of high-yield peanut varieties and use of winged bean, flowers, tender pods, mature seeds and tubers as a routine in every subsistence garden and household
— Utilization of tender tapioc leaves in the diet of the people
— Cooking and consumption of kaukau in the unpeeled form and steaming and roasting of unpeeled sweet potato instead of boiling
— To encourage steaming and roasting and teach the people how to make bamboo steamers and use them for their cooking (Fig. 142)

Fig. 142 Construction and use of a bamboo steamer

— Get the cooperation of the people and construct communal single drum ovens to help them to roast their food or to steam their vegetables (Fig. 143).

Subsistence agriculture and aspects of ecology and conservation in Papua New Guinea

Introduction

Nine years since Independence and 100 years since colonial rule was initiated, development in Papua New Guinea is now beginning to accelerate on various levels.

The copper and gold mining project at Ok Tedi is under way. The Gogol timber area in the Madang Province with more than 67,000 ha of clearfell has opened up, followed by the approval of others like the Kumusi Timber Project in the Northern Province with 85,000 ha and the Kapuluk Timber Project in West New Britain with 181,000 ha. The population has increased to a total of 3,006,799 people in Papua New Guinea and brought problems of land shortages to subsistence farmers as well as pressure on social services. One of the most affected provinces is the Morobe Province with a population of 310,526 people.

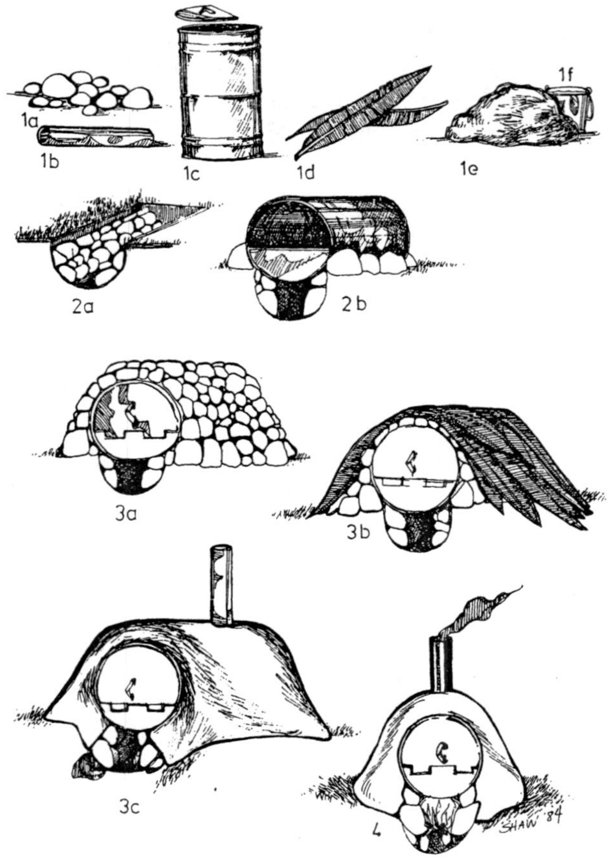

Fig. 143 Construction of a drum oven

1 Material necessary: 1a Stones, 1b Chimney, 1c Precleaned drum (the drum must be burned to clean it from all possible detergents left)

1d Banana leaves, 1e clay, 1f water

2a Construction of the fire place underneath the drum

2b Placing of the drum

3a Stone layer around the drum and proper cutting of the door site

3b Banana leave layer around the drum

3c Clay layer and placement of the chimney and fixing of the door of the drum oven made out of the cut off drum lid

4 Drum oven ready for use

Papua New Guinea is fortunate to be a country of outstanding natural beauty and biological interest. Its geological history has resulted not only in rich natural deposits, but in a flora and fauna that are amongst the most interesting to be found anywhere in the world (Paijmans, 1976; Gressitt, 1981). The great majority of the population depend directly upon this flora and fauna for virtually every necessity of their day-to-day lives (Moranta et. al., 1982).

Government policy

The Papua New Guinean government has the intention to rapidly develop the natural resources of the country. This is considered necessary for several reasons, but due to the fact that the country has only a small industrial base, the only way to get the needed income is by exploiting the natural resources.

The fourth of the five National Goals layed down in the Constitution of Papua New Guinea was concerned with natural resources and environment:

"We declare our fourth National Goal to be for Papua New Guinea's natural resources and environment to be conserved and used for the collective of future generations.

We accordingly call for:

1. Wise use to be made of our natural resources and the environments in and on the land or seabed, under the land, and in the air, in the interest of our development and in trust for future generations; and

2. The conservation and replenishment, for the benefit of ourselves and prosperity, of the environment and its sacred, scenic and historical qualities; and

3. All necessary steps to be taken to give adequate protection to our valued birds, animals, fish, insects, plants, trees."

In 1976 the Parliament endorsed an Environment and Conservation Policy as laid down by the Office of Environment and Conservation (OEC). All three divisions of the OEC —

Environment, National Parks and Wildlife—have been seriously affected by the 1982 budget proposal causing the discontinuation of a variety of important services (Table 7).

Table 7. Effects of the 1982 budget proposal on the programmes of the Office of Environment and Conservation (Chambers, 1983)

Office of Environment and Conservation		
Environment Division	National Parks Service	Wildlife Division
Planning and Assessment (1)	Management and Operations (1)	Wildlife Management (1)
Environmental Resources (1)	Resource Investigations (3)	Research and Surveys (1)
Environmental Protection (1)	Education and Interpretation (3)	Crocodile Project (2)
Environmental Education (3)		Insect Farming (2) Law Enforcement (3)
Population Programme (1)		Provincial Wildlife (3) Wildlife Education (3)
		Wildlife Library (3) Wildlife Publications (3) Dudong Project (3) Turtle Project (3)

1: Continuation reviewed; 2: Transferred to the Department of Primary Industries; 3: Discontinued

Some of the results of disbanding, for example, the Law Enforcement Section, is that the door is opened for gross abuse of the country's fauna and flora protection laws. With the present market value of a pair of Raggian Birds of Paradise at Kina 15,000 and a Queen Alexandra butterfly at Kina 500, such speculations are not unrealistic.

The transfer of the responsibility of the insect and crocodile farming to the Department of Primary Industry has put a very specific responsibility on the shoulders of people not very familiar with the very specialized nature of these projects. It is feared that, besides threatening the future of rare insects, it will be a tragedy for a variety of small village — level businesses the agency helped to establish.

Such rural projects show the way that development in rural areas can proceed.

Discussion and conclusions

The aims and the goals of the 4th National Goal have been jeopardized and it is therefore of great importance that all people involved and interested in the development of this country should do their part to prevent destruction of the environment. The best conservation schemes evolve from the bottom up and are not imposed from the top down. In a country where 87% of the population lives in rural areas and where their food, medicine, the raw material for their houses and shelters, tools, canoes, weapons, and clothing is obtained directly from the environment, any further deterioration of the natural environment will have a great impact on subsistence lifestyles.

If no adequate handling of the environment can be placed into the hearts and brains of the people, severe changes in lifestyle and extreme deterioration will be the result. Landslides and floods, pollution and an increase in water-borne diseases will only be the first steps (Fig. 144).

Everybody can do his part to prevent these inevitable developments and to avoid the syndrome which pervaded in the early history of many so called developed countries. For example, if a man cut down one tree or two he ran the risk of being a vandal; if he cut down a few hundred trees he may have become a successful timber man; if he destroyed millions of trees he was acclaimed a pioneer.

Fig. 144 The possible effects of inappropriate use of forests
If the forests are all gone then the topsoil will be eroded away leaving only the bare rocks. As a result floods will occur in the lowlands and heavy siltation of the rivers will lead to the destruction of the spawning sites of fishes.

Let us avoid this sort of development by showing everyone where he can contribute to conserve the unique environment of Papua New Guinea and at the same time get all the necessary basic needs. The forests should not be regarded as a hindrance, but as the necessary area and resource center for a brighter prospect in subsistence agriculture.

Aspects of subsistence agriculture and education in Papua New Guinea

Introduction

The problem of schooling in Papua New Guinea, as in many other developing nations is that educational provision has to cater to a dual society: a traditional more rural-oriented and a modern, more urban-oriented. While the former is by far the characteristic life situation in the country, adjusting from subsistence agriculture to a partly commercial economy, most of the government schools are geared to the latter by preparing students mainly for higher-level manpower needs, necessary for the development of the country at the national level.

There is no doubt that the diversity of living conditions in Papua New Guinea demands the development of partly different school systems and curricula, if the goal of an environment-oriented schooling will be achieved. The diversity of the physical conditions, which is to a considerable extent responsible for different socio-economic environmental regions, hampers the development of a uniform curriculum for the country.

There is therefore a widely recognized controversy as to whether the topics of instruction have to differ between regions in order to achieve the aim of an environmental-oriented education. Environmental-oriented education means a more village-oriented approach in education in its best sense. If we realize that 2/3 of the Gross National Product (GNP) is produced by the agricultural sector and that in 1980 about 2,613,000 people out of a total 3,006,799 citizens (about 87%) lived in villages, then a more environmental-oriented education is justified.

Government policy and parents expectations

After independence in 1975 the National Government drafted an "Eight Point Plan" as a guideline for the socio-economic development of the country. This plan aims at promoting self-reliance, growth of small scale artisan village industry, equal distribution of wealth and equal participation of women and men in building the nation.

How can schools and other educational institutions contribute to its implementation? In questioning parents about their expectations they generally expressed the opinion that the school teach the pupils how they can assist in a direct material improvement of village life after their graduation (Kemelfield, 1976). In particular, the pupils are expected to learn to earn money through agricultural and commercial activities. But very often parents realized, when their children had finished primary school, that "their educational knowledge is just too little to give any help to the village community" (Kemelfield, 1976).

A growing number of parents are becoming aware of the fact that most of the school graduates are not able to go to high school or find any employment in towns.

The present school system in Papua New Guinea

There are two main school systems in Papua New Guinea, the Government School System and the Vernacular School System, mainly run by the churches.

Very often extension officers and agronomists working in the rural areas have to cooperate with semi-educated school leavers and drop outs of the 6th or 10th grade.

It is therefore of great importance to realize that until today all these pupils, which normally get no chance to receive any proper formal education, are bound to go back to their village or stay in their village without having received any village-oriented or environment-oriented education. The schools have very often created "job seekers" and not "job makers" (Bude, 1973), unemployment of school leavers, white collar workers, a drift to town, uncontrolled urbanization, and unwanted change of traditional way of life and values.

Reform and development necessities of the environment-oriented education

A worldwide recognized method of reforming and adjusting current educational practice in developing countries has been developed by Paulo Freire (Freire, 1972). He worked out his findings in South America and his method is carried out in three steps:

1. Phase of investigation : Analysis of the life situation in a specific area.
2. Phase of codification : Topics of a problematic nature of the life situation of the people are collected and codified in key-words, representing keytopics.
3. Phase of decodification : Discussion of the keytopics with attendants of projects and courses in the villages or schools.

To adopt this method to the Papua New Guinea situation, three regions were under investigation during 1975 – 1977 (Heinrich et. al., 1975; Gintausu et el., 1977). The analysis of the determinants of the life situation in the Central Highlands (C.H.R.), the Markham-Ramu Region (M.R.R.) and the Coastal-Island Region (C.I.R.) of New Guinea revealed the following keytopics (Osmers, 1977):

C.H.R.	M.R.R.	C.I.R.
Coffee	Banana	Seafood
Pig	Cattle	Copra
Sweet Potato	Coconut	Coconut
Tradestore	Traffic	Taro

With these findings in mind the education and any approach used to improve the village situation in these regions should start with these most important determinants of the life situation in the various regions. For example, in the Highlands the discussion and instruction of the keytopic, pig, should take place in the classroom and in a pig husbandry project, covering the following topics:

THEORETICAL CLASSROOM INSTRUCTION	PROJECT INSTRUCTION
1. Traditional pig husbandry.	1. Measuring of land plots and preparing of gardens.
2. Traditional tribal stories of the pig.	2. Nutrition of pigs and planting of crops.
3. The pig as nutritive animal.	3. Designing and drawing and building of a pig house.
4. The pig exchange ceremony.	4. Preparing and buying of pig feed and cleaning of the pig house.
5. Prohibition of the participation of Lutheran adherents in pig exchange ceremonies.	5. Composting or use of organic manure and planting of new crops.
6. Decision making in Papua New Guinea.	6. Care of pigs and breeding of pigs.
7. Modern pig husbandry.	7. Pig diseases.
8. Animal husbandry in other countries.	8. Butchering of pigs.
9. Budget calculation and book-keeping.	9. Buying and selling of pigs.

The other keytopics should be taught in a similar way.

In mixed areas, like the Bulolo-Wau Area, which has combination of coastal and highland characteristics, the education could be adapted by teaching the pupils the following topics as first priority:

THEORETICAL CLASSROOM INSTRUCTION	PROJECT INSTRUCTIONS
Coffee Program	Coffee Project
Tradestore Program	Tradestore Project
Coconut Program	Handicraft Project
Sweet Potato Program	Poultry and vegetable project
Taro Program	Fishpond Project
	Bakery Project

Included in the coffee project would be, location, appearance, cultivation, processing, and selling.

By starting with and stressing the main topics of the life situation of the village people in a given area, the improvement of the entire life situation can be achieved because the people are especially interested in promoting these topics which are directly related to their own life situation. By this means the education system would contribute to a more balanced education and economical means to reach the goal of self-reliance, stated as one of the National Goals in Papua New Guinea.

Useful addresses

1. Department of Primary Industry
   P.O. Box 417 Konedobu Papua New Guinea

2. ELC-PNG Viles Teknik Mainyanda
   P.O. Box 87 Bulolo Papua New Guinea

3. Liklik Buk Information Centre
   P.O. Box 1920 Lae Papua New Guinea

4. Morobe Womens Association
   P.O. Box 1327 Lae Papua New Guinea

5. SPATF
   P.O. Box 6937 Boroko Papua New Guinea

6. ATDI University of Technology Lae
   Free Mail Bag Lae Papua New Guinea

Literature

Bary, R.G., 1978. Aspects of the precipitation characteristics of the New Guinea mountains. J. Trop. Geogr. 47, 13–30.

Bude, U., 1973. Ruralisation — Information — Participation. In: Entwicklung und Zusammenarbeit 6, Bad Godesberg, West Germany.

Chambers, M.R., 1983. Environmental Programmes and Policies in Papua New Guinea since Independence. In: Environmental Education Series No. 2, UNPNG, Port Moresby.

Charles, A.E., 1975. Shifting cultivation and food crop production. In: Proc. Papua New Guinea Food Crop Conf., DPI, Konedobo.

Cox, E., 1979. Gavien and Bagi: Rubber/Profit versus People/Community. In: C.A. Valentine and B.L. Valentine (ed.), 1979: Going through changes: Villages, Settlers and Development in Papua New Guinea. JPNGS. Port Moresby.

Dalpadado, V.E., H. Sakulas, 1982. The application of work study to improve subsistence agriculture systems in the developing countries. E.F. Schumacher Memorial Conference. Univ. of Sydney, Sydney, Australia.

Department of Primary Industries, 1984. Farming Notes No. 1–30. Publ. Section DPI, P.O. Box 417, Konedobo, Papua New Guinea.

Dronia, H., 1983. Bodentemperaturmessungen in tropischen Gebirgen. Erdkunde Bd. 37, 292-295.

Freire, P., 1972. Conscientization for Liberation. Washington. CICOP Public., Washington D.C., USA.

Gagne, W., 1979. Natural Control of insect pests in an organic garden. South Pacific Commission Workshop on Biological Control, Noumea. Leaflet No. 3, Wau Ecology Institute, P.O. Box 77, Wau, Papua New Guinea.

Gagne, W., 1980. The transformation and intensification of shifting agriculture: Past and present conservation practices. In: L. Moranta, J. Pernetta, W. Heaney (ed.), 1980. Traditional Conservation in PNG: Implications for today. Monograph 16, JASER, Port Moresby.

Gressitt, J.L. (ed.), 1982. Biogeography and Ecology of New Guinea. Monographiae Biologicae Vol. 42. Junk Publ., London.

Hale, P.R., B.D. Williams, 1979. Liklik Book. Liklik Book Information Centre, P.O. Box 1920, Lae, Papua New Guinea.

Kalma, J.D., 1978. Solar radiation over New Guinea and adjacent islands. Austr. Met. Mag. 20. 116–127.

Kemelfield, G. 1976. Planting the seed: A proposal for community-based education. In: Thomas, E.B. (ed.), 1976.

Lamb, K.P., Gressitt, J.L. (ed.). 1976. Ecology and conservation in Papua New Guinea. Wau Ecology Institute Pamphlet No. 2, Wau Ecology Institute, P.O. Box 77, Wau, Papua New Guinea.

Lamb, K.P., Johnston, A.D., 1976. Some insect pests of Papua New Guinea, Dep. of Biology, Univ. Papua New Guinea, Port Moresby.

Leonard, D., 1967. Soils, Crops and Fertilizers Use. A what, how, and why guide. Program and Training Journal Reprint Series No. R8, Peace Corps, 806 Connecticut Avenue, Washington D.C., 20525, USA.

Mc Alpine, J.R., G. Kreig, K. Short, 1975. Climatic tables for Papua New Guinea. CSIRO. Div. of Land Use Res., Techn. Pap. 37.177p.

Moranta, L., J. Pernetta, L. Heraney, 1982. Traditional Conservation in Papua New Guinea: Implications for today. JASER Publ., Port Moresby.

Osmers, D., 1977. The vernacular education program of the evangelical Lutheran Church of PNG. Mainyanda. Project Report.

Pajimans, K., 1976. New Guinea Vegetation. Elsevier Publication, Amsterdam, The Netherlands.

Roger, B., Yepsen, J. (ed.), 1976. Organic Plant Protection. A comprehensive reference on controlling insects and diseases in the garden, orchard and yard without using chemicals. Rodale Press Inc. Emmaus, Pennsylvania 18049, USA.

Vergara, N.T., 1982. Sustained outputs from Legume-Tree-Based-Agroforestry-Systems. New Directions in Agroforestry: The potential of tropical legume trees. Environment and Policy Institute, East-West Center, Honolulu, Hawaii, USA.

Vergara, N.T., 1982. Improving agroforestry in the Asia-Pacific-Tropics. New directions in agroforestry: The potential of tropical Legume trees. East-West Center and Nat. Univ., Honolulu, Hawaii, USA.

Williams, A.R., 1982. Biological Conservation techniques to help control soil erosion. In: Proc. 2nd PNG Food Crop Conf., 1982, DPI, Publ. Sec., Konedobo, Papua New Guinea.

INDEX

PUBLICATIONS OF WAU ECOLOGY INSTITUTE

HANDBOOKS

Handbook of common New Guinea frogs. J.I. Menzies. 75 p; 12 col. plates

Handbook of common New Guinea beetles. J.L. Gressitt and R.W. Hornabrook. 87 p; 4 col. plates and other illustr.

Guide to biological terms in Melanesian Pidgin. M. Simon. 115 p; illustrated

Guide to montane birds of Northeast New Guinea. B. Beehler. 157 p; 5 col. plates; 5 black and white plates; other illustr.

Guide to Mt. Kaindi: Background to montane New Guinea ecology. J.L. Gressitt and N. Nadkarni. 135 p; many illustr.

Handbook of New Guinea rodents. J. Menzies and E. Dennis. 68 p; 8 col. plates

Reptiles of the Solomon Islands. M. McCoy. 80 p; 8 col. plates

Birds of North Solomons. D. Hadden. 107 p; 24 col. plates

Birds of New Guinea. B. Beehler, T.K. Pratt and D.A. Zimmerman. Princeton Univ. Press. 293 p; 55 plates

PAMPHLET

Ecology and Conservation in Papua New Guinea. K.P. Lamb and J.L. Gressitt (eds.)

CPSIA information can be obtained at www.ICGtesting.com
Printed in the USA
BVOW04s1809220914

367644BV00007B/64/P

9 789980 945761